Samuli Niiranen and Andre Ribeiro

Information Processing and Biological Systems

Intelligent Systems Reference Library, Volume 11

Editors-in-Chief

Prof. Janusz Kacprzyk
Systems Research Institute
Polish Academy of Sciences
ul. Newelska 6
01-447 Warsaw
Poland
E-mail: kacprzyk@ibspan.waw.pl

Prof. Lakhmi C. Jain
University of South Australia
Adelaide
Mawson Lakes Campus
South Australia 5095
Australia
E-mail: Lakhmi.jain@unisa.edu.au

Further volumes of this series can be found on our homepage: springer.com

Vol. 1. Christine L. Mumford and Lakhmi C. Jain (Eds.)
Computational Intelligence: Collaboration, Fusion and Emergence, 2009
ISBN 978-3-642-01798-8

Vol. 2. Yuehui Chen and Ajith Abraham
Tree-Structure Based Hybrid Computational Intelligence, 2009
ISBN 978-3-642-04738-1

Vol. 3. Anthony Finn and Steve Scheding
Developments and Challenges for Autonomous Unmanned Vehicles, 2010
ISBN 978-3-642-10703-0

Vol. 4. Lakhmi C. Jain and Chee Peng Lim (Eds.)
Handbook on Decision Making: Techniques and Applications, 2010
ISBN 978-3-642-13638-2

Vol. 5. George A. Anastassiou
Intelligent Mathematics: Computational Analysis, 2010
ISBN 978-3-642-17097-3

Vol. 6. Ludmila Dymowa
Soft Computing in Economics and Finance, 2011
ISBN 978-3-642-17718-7

Vol. 7. Gerasimos G. Rigatos
Modelling and Control for Intelligent Industrial Systems, 2011
ISBN 978-3-642-17874-0

Vol. 8. Edward H.Y. Lim, James N.K. Liu, and Raymond S.T. Lee
Knowledge Seeker – Ontology Modelling for Information Search and Management, 2011
ISBN 978-3-642-17915-0

Vol. 9. Menahem Friedman and Abraham Kandel
Calculus Light, 2011
ISBN 978-3-642-17847-4

Vol. 10. Andreas Tolk and Lakhmi C. Jain
Intelligence-Based Systems Engineering, 2011
ISBN 978-3-642-17930-3

Vol. 11. Samuli Niiranen and Andre Ribeiro (Eds.)
Information Processing and Biological Systems, 2011
ISBN 978-3-642-19620-1

Samuli Niiranen and Andre Ribeiro

Information Processing and Biological Systems

Springer

Dr. Samuli Niiranen
Department of Signal Processing
Tampere University of Technology
POB 553
33101 Tampere
Finland
E-mail: samuli.niiranen@tut.fi

Dr. Andre Ribeiro
Department of Signal Processing
Tampere University of Technology
POB 553
33101 Tampere
Finland
E-mail: andre.ribeiro@tut.fi

ISBN 978-3-642-19620-1   e-ISBN 978-3-642-19621-8

DOI 10.1007/978-3-642-19621-8

Intelligent Systems Reference Library   ISSN 1868-4394

Library of Congress Control Number: 2011923210

© 2011 Springer-Verlag Berlin Heidelberg

This work is subject to copyright. All rights are reserved, whether the whole or part of the material is concerned, specifically the rights of translation, reprinting, reuse of illustrations, recitation, broadcasting, reproduction on microfilm or in any other way, and storage in data banks. Duplication of this publication or parts thereof is permitted only under the provisions of the German Copyright Law of September 9, 1965, in its current version, and permission for use must always be obtained from Springer. Violations are liable to prosecution under the German Copyright Law.

The use of general descriptive names, registered names, trademarks, etc. in this publication does not imply, even in the absence of a specific statement, that such names are exempt from the relevant protective laws and regulations and therefore free for general use.

Typeset & Cover Design: Scientific Publishing Services Pvt. Ltd., Chennai, India.

Printed on acid-free paper

9 8 7 6 5 4 3 2 1

springer.com

Preface

What is information? What is its role in biology? Late in this "Information Technology" era, one would think that we knew the answers to these issues. But, in fact, we do not. This book, edited by Samuli Niiranen and Andre Ribeiro, is a fine effort to explore these issues.

For readers not familiar with the debate, I will outline the two main theories of information and briefly discuss them.

In 1948 Claude Shannon invented his information theory. Shannon was concerned about sending signals down a telephone channel subject to noise. He rightly realized that for this engineering problem, the "meaning" of the information sent down the channel was irrelevant.

Shannon's first idea was that the simplest "unit" of information was a "yes" versus "no" choice, which could be encoded by a 1 or 0 symbol. Then a message sent down a channel would be some string of 1 and 0 symbols.

Brilliantly, Shannon sought to define the amount of information in such binary symbol strings. To do so, he imagined a source that emitted symbol strings. To explain this idea, imagine that each symbol string is length N. Then the "source" might contain many copies of one symbols string, called a "message", and few copies of other symbol strings. Shannon wanted an additive measure of information. He invented a mathematical formula to accomplish this. He considered the number of copies of each message, message "i", divided by the total number of messages in the source, say M, to define the probability Pi, that the source would "emit" that message, then made an additive theory of the amount of information in the source by taking the logarithm of Pi to a logarithmic base that he chose to be base 2. Then he calculated the average value of Log Pi for all messages in the source, and made this a positive number: - sum PiLogPi.

An amusing and perhaps true story is that Shannon asked von Neumann, the famous mathematician, what to call his measure of the information in the Source. "Call it entropy", goes the story, "No one knows what entropy is". Whether true of false as a story, this mathematical expression is called the "Entropy of the information source".

Shannon's mathematical expression is essentially the same as Boltzmann's expression in statistical mechanics for the entropy of a physical system.

Then Shannon imagined his 1 and 0 binary symbols, now called "bits", being sent down an information channel. But the channel might be noisy. He then invented the idea of an error correcting code. The simplest is this: encode each binary symbol, or bit, as a triplet of identical bits, so the symbol string 1011 becomes 111000111111. Then his idea was that a "decoder" at the far end of the channel could decode this encoded message and error correct by what is called "the majority rule". If the decoder received the symbol string 101000111110 it would examine each triplet and guess that the first triplet 101 was really 111 and that noise had degraded, or flipped the middle 1 to 0. These ideas led to the generation of wonderful error correcting codes and Shannon's theorems about the carrying capacity of his channel in the face of noise.

But I ask you to notice three things about Shannon's theory. First, by design, it has no "semantics". We have no idea or definition of what the 1 and 0 symbols "mean". This issue was irrelevant to his engineering problem. Second, Shannon assumes that the "decoder" can have indefinite computational power. It can decode any arbitrary encoding. This assumption is entirely unrealistic. In this book, primary examples of information processing systems are cells and genetic regulatory networks. Such causal systems have no decoder of indefinite computational power. This renders Shannon's Channel theorems questionable or moot. For the deeper question is, how causal systems can coordinate their behaviors when no parts of any of the systems have indefinite computational power. This book seeks answers to this fundamental question.

Third, Shannon supplies a measure of how much information the source has, his Source entropy. But Shannon never tells us what information "is". To this day, computer scientists give a diversity of answers to this puzzling question. I will suggest my own try below. This suggestion is implicit in the work described in this book.

The second theory of information we have is due to Kolmogorov. He defined the information content of a symbol string, say a binary symbol string, as the length of the shortest computer program on a universal computer that could produce the symbol string. Thus, consider the symbol string (1111111111). The shortest program for this is: "Print 1 ten times". Now consider (10101010101). The shortest program is: Print "10 five times". Notice that this measure takes advantage of the redundancy in the symbol string. Kolmogorov realized and proved that for a completely random symbol string, the shortest program would need to be proportional to the length of the random symbol string. This idea has led to brilliant work. For example, consider a random symbol string length N. Concatenate it with itself to create a symbol string length 2N. Now "compress" the 2N string to get rid of all redundancy. An ideal compressor will yield a compressed symbol string length N. But now take two different random symbol strings length N. Concatenate them and try to compress them. You cannot compress the two strings to a length less than 2N. So the extent to which you can compress two symbol strings is a measure of how similar they are. Indeed it is a measure of the mutual information between the strings. Conversely, the extent to which the two strings cannot be compressed is a measure of how different they are. This can be used to construct a universal measure of the "distance" between symbol strings, the "normalized compression distance" (NCD). This NCD is now being put to interesting use.

Notice about Kolmogorov's measure that he does not tell us what information "is".

The fact that neither Shannon nor Kolmogorov tell us what information "is", has left us with the impression that information somehow floats free, some abstract something and we do not know what it is.

This book concerns information either in cells or language. I want to close this prologue with my own try at deepening the idea of what information "is". It is concurrent with the contents of this book, as I'll sketch below.

In What is Life, published in 1943, Schrödinger brilliantly guesses that the gene, estimated to have only a few hundred atoms, must be a "solid" held together via quantum mechanical chemical bonds, the true source of order in organisms, instead of the familiar classical statistical mechanics of ink diffusing in a petri plate. Then he says, of the gene, "It will not be a regular crystal, they are dull. It will be an aperiodic crystal containing a microcode for the organism." This is all Schrödinger tells us. He does not elaborate on what he means.

I'm going to give my own interpretation. I begin with physical work. Work is defined as "force acting through a distance", for example, a hockey stick accelerating a hockey puck. But P. W. Atkins in his book on the Second Law says "Work is a thing. It is the constrained release of energy into a few degrees of freedom." For example, consider a cylinder and piston with the hot working gas in the head of the cylinder. The random motions of the hot gas exert pressure on the piston which moves in translational motion down the piston in the first part of the power stroke of a heat engine.

This is precisely the release of energy into a few degrees of freedom, the translational motion of the piston down the cylinder. Thus work is done.

But what are the constraints? The cylinder, the piston, the location of the piston inside the cylinder and, of course, the working gas constrained to the head of the cylinder.

Now the physicist will put in these constraints as fixed and moving boundary conditions.

But how did these constraints come to exist in the universe? Well, it took work to construct the cylinder and piston and locate the piston and gas inside the cylinder. So we come to something not in the physics books: It (typically) takes work to construct Constraints on the release of energy into a few degrees of freedom that constitutes Work. No work, no constraints. No constraints, no work. We lack a theory of this. Hints of it are in this book as I'll note below.

Now back to my interpretation of what Schrödinger was getting at. I think the aperiodic crystal, for example the DNA molecule with its arbitrary sequence of bases, contains myriad broken symmetries that are the microconstraints that allows energy to be released on constrained ways and constitute a myriad of diverse work processes.

I want to suggest that this yields the start of an embodied sense of information. Here, the meaning of information is the work process that is enabled. A measure of how much will require a measure of how organized this work process is, and how diverse the set of work processes are. I think these measures can be constructed.

These suggestions tie to this book. In D. Cloud's initial chapter, he rightly notes that in cells, information involves ignoring details of molecular motions and features and the cell treating a DNA sequence as "the same sequence" despite how it wiggles in water. Cloud's insight is an example of throwing away some aspects of the detailed molecular behavior and the cell's use of an equilvelence class of detailed configurations of the DNA encoding an RNA to transcribe the same RNA. In a way that is not quite clear yet, Cloud's "ignoring" creates an equivelence class of detailed sequence configurations that together constitute the constraints on the RNA polymerase to transcribe the gene into RNA. This allows the specific work of transcription to happen.

Now in much of the book, dynamical models of gene regulation and genetic regulatory networks are discussed. Work is not at the heart of these discussions, but is implicit in them. This is true because transcription and translation, as noted above, is the constrained release of energy into a few degrees of freedom, and the genetic regulatory network's dynamics is the further constrained release of energy into few degrees of freedom. Thus all the discussions in this book about network dynamics links causal networks implicitly with work processes that are the constrained release of energy. I suspect that this gets at what information in cells "is".

In the final chapter, S. Niiranen discusses language. But here too in a still poorly articulated way, constraints are hiding. Consider the child game of hopscotch. There are lines of chalk on the sidewalk and the rules of hopscotch that jointly are enabling constraints that organize the activities of children around the world to play hopscotch. No constraints, no organized activities. Language has its constraints in its syntax, let alone its semantic meanings which, I believe require conscious experiences and an analysis of reference.

This is a fine book. In examining cell dynamical systems and genetic regulatory networks that are models of causal classical physical systems that embody the constrained release of energy, and an analysis of language, the book moves us toward what will someday become a much richer understanding of information and its role in the unfolding complexity of the universe.

November 25, 2010   Stuart A. Kauffman

Contents

Introduction 1
Samuli Niiranen, Andre Ribeiro

Biological Information and Natural Selection 9
Daniel Cloud

Swarm-Based Simulations for Immunobiology: What Can
Agent-Based Models Teach Us about the Immune System? 29
Christian Jacob, Vladimir Sarpe, Carey Gingras, Rolando Pajon Feyt

Biological Limits of Hand Preference Learning Hiding Behind
the Genes 65
Fred G. Biddle, Brenda A. Eales

Stochastic Gene Expression and the Processing and
Propagation of Noisy Signals in Genetic Networks 89
Daniel A. Charlebois, Theodore J. Perkins, Mads Kaern

Boolean Threshold Networks: Virtues and Limitations for
Biological Modeling... 113
Jorge G.T. Zañudo, Maximino Aldana, Gustavo Martínez-Mekler

Structure-Dynamics Relationships in Biological Networks....... 153
*Juha Kesseli, Lauri Hahne, Olli Yli-Harja, Ilya Shmulevich,
Matti Nykter*

Large-Scale Statistical Inference of Gene Regulatory
Networks: Local Network-Based Measures 179
Frank Emmert-Streib

Information Propagation in the Long-Term Behavior of Gene
Regulatory Networks 195
Andre S. Ribeiro, Jason Lloyd-Price

Natural Language and Biological Information Processing 219
Samuli Niiranen, Jari Yli-Hietanen, Olli Yli-Harja

Author Index 229

Introduction

Samuli Niiranen and Andre Ribeiro

Computational Systems Biology Research Group, Department of Signal Processing, Tampere University of Technology, Tampere, Finland

Abstract. Information propagation and processing pervades biological systems. While it is known that there is information processing and propagation at the various levels of detail, such as within gene regulatory networks, within chemical pathways in cells, between cells in a tissue, and between organisms of a population, how this occurs and how information is processed in living systems is still poorly understood. For example, it is not yet entirely understood how gene networks process the information contained in changes in the temporal level of an mRNA so as to distinguish a stochastic fluctuation from a message containing valuable information regarding another genes state. In recent decades, information theoretical methods and other tools, originally developed in the context of engineering and natural sciences, have been applied to study various aspects of diverse biological processes and systems. This book includes chapters on how the processing of information in biological processes is currently understood in various topics, ranging from information processing and propagation in gene regulatory networks to information processing in biological olfaction and natural language. We aim at presenting an overview on the state-of-art on how information processing relates to biological systems and the opinion of current leaders in the field on future directions of research.

1 Information in Biology

Biological systems are constantly engaged in activities that can be perceived in terms of information processing or informational representation. Some obvious examples are perception, cognition, and language use. During the last few decades, informational concepts have been applied to a wide range of biological processes. [1]

For many working in biology, the most basic processes of biological systems should now be understood in terms of the expression of information, the execution of programs, as well as the interpretation of codes [1]. Thus, concepts from information theory have been applied to the study of biological processes. However, a major problem exists. In the definition of information proposed by Claude Shannon [2], there is no context, that is, information is not quantifiable as a context dependent quantity, while in biological systems information is always context dependent. Great care is thus necessary when applying the concepts from information theory into biological settings.

The use of concepts from information theory has acquired a prominent role in various fields in biology, particularly in genetics, developmental biology and evolutionary theory, the latter ones especially where they border on genetics. One distinctive use of informational concepts has been in describing the relations between genes and the various cellular structures and processes that genes' activity are known to affect. For many geneticists, the causal role of genes should be grounded in them carrying information via the products of their expression to other genes, not only to other genes, but to everywhere in the cell. [3]

The concept of "gene" is, in modern theories, crucial to both explain organisms' functioning and the inter-generational inheritance of characteristics, where it is seen as the "information carrier".

Currently these two explanatory roles are anchored in a set of well-established facts about the role of DNA and RNA on protein synthesis in cells, summarized in the familiar chart representing the "genetic code", mapping DNA base triplets to amino acids. However, the use of informational concepts in biology pre-dates even a rudimentary understanding of these mechanisms [4]. Importantly, contemporary applications of informational concepts extend far beyond this relatively straightforward case of specification of protein molecules by DNA. These include [1]:

1. The description of phenotypic traits of organisms, including complex behavioral traits, as specified or coded by information contained in genes or in the brain.
2. The description of various causal processes within cells, and possibly of the whole-organism developmental sequence, in terms of a program stored in genes.
3. Treatment of the transmission of genes, and sometimes other inherited structures, as an inter-generational propagation of information.
4. Treatment of genes themselves, in the context of evolutionary theory, in some sense as constituted of information.

Traditionally, information and concepts from information theory are interpreted within a biological setting as follows: informational connections between events or variables involve nothing but ordinary correlations, possibly related to underlying physical causation. In this sense, a signal carries information about a source if we can predict the state of the source from the signal. This view of information is derived directly from Claude Shannon's work [2], where he proposed the use of the concept of information to quantify facts about contingency and correlation, initially for use in digital communication technology. This information-theoretical approach has been applied in biology as part of the attempt to use computational data analysis tools on biological problems as we will see in this book. Biologists use this sense of information in a description of gene action or other processes of biological systems, in an attempt to adopt a quantitative framework for describing ordinary correlations or causal connections [1]. Note that, in this perspective, biological processes are not being explained in terms of the use or manipulation of information.

A currently contentious question is whether biology needs another, richer concept of information in addition to "causal" Shannon information. Information in this sense is sometimes referred to as "semantic" or "intentional" information. A simple example how the problematic regarding the use of the concept of information as proposed by Shannon in a biological setting goes as follows: in order to quantify the amount of information contained in a message, one needs to know the entire state space of possibilities, so as to compute the probability of occurrence of the content received in the message. Unfortunately, in most realistic scenarios, such state space is unknown or even transcomputational, thus the probability value is unknown. One possibility is that prior knowledge derived from previous similar occurrences is used to set a value to the probability of the present message, but even this is somewhat problematic regarding how such quantification is actually accomplished. For these reasons, exact quantification of information content in biological contexts has so far been restricted to a very restricted set of biological problems, usually when it is possible to define a simplified model of the system under study. Here, we present a number of these studies, some of which deal with information as an explicit variable, others focusing on the study of the dynamics of systems where information propagation plays a major role.

This book includes chapters both on how the processing of information in biological processes is currently understood from low-level cellular and tissue-level processes to the high-level functions of cognition and language acquisition. We aim at presenting an overview on the state-of-art on how information processing relates to biological systems and to present the opinion of current leaders in the field on future directions of research.

2 Chapters Included in This Book

The remainder of the book consists of ten chapters and a preface by Stuart. A. Kauffman.

In Chapter 1, Dr. D. Cloud discusses biological information in the context of natural selection. D. Cloud discusses the concept of biological information and its relation with the concept of complexity. Relevantly, Cloud distinguishes between complexity of natural processes and complexity of biological systems.

It is proposed that biological information in cells is a set of simplifying conventions used to coordinate the division of labor. It is also discussed why natural selection might produce such systems of conventions. A simple formal model of the relationship between natural selection, complexity and information is proposed and then its applicability in the context of adaptive landscapes is explored.

An interesting question is raised and discussed, regarding how evolution may lead to the emergence of complexity.

In the following chapter, Prof. C. Jacob and colleagues present us both a review of previous results, as well as recent ones, of a long ongoing work towards developing swarm-based models and simulations of biological systems. Here, they mostly focus of models for immunobiology.

Jacob and colleagues explain how swarm intelligence techniques can used to explore key aspects of the human immune system. They present us three models and the results of their virtual simulations. In these models, Immune system cells and related entities (viruses, bacteria, cytokines) are represented as virtual agents inside 3-dimensional, decentralized compartments that represent primary and secondary lymphoid organs as well as vascular and lymphatic vessels.

Using these models it is shown how targeted responses of the immune system emerge as by-products of the collective interactions between the agents and the environment. In particular, they show simulation results for clonal selection in combination with primary and secondary collective responses after viral infection. They model, simulate, and visualize key response patterns due to bacterial infections. Finally, they consider the complement system. In the end, they discuss how in-silico experiments are essential for developing hierarchical whole-body simulations of the immune system, both for educational and research purposes.

In Chapter 3, Prof. F. Biddle and B. Eales discuss their latest findings on the process of learning and memory, as well as on the limits of learning. A detailed description is made on the research of the genetics of learning and memory in hand preference of laboratory mice. The research laid out focuses on better understanding of how mice acquire a bias in paw preference, how it varies between different mice genotypes, and whether hand preference is a constitutive or an adaptive behavior.

Relevantly, they describe a new model able to explain the nature, as well as predict, the dynamics of the acquisition of bias in paw preference, both at the individual level, as well as at the population level, representative of the genotype.

It is reported how patterns of paw-preference scores of inbred mouse strains contain more information than what can be inferred from means and variances of the quantitative behavioral scores. From these, it is explained how qualitative differences in the dynamic patterns of paw preference revealed the learning and memory process in the mouse behavior and how the process is genetically regulated.

In Chapter 4, D. Charlebois and colleagues report recent results of their studies of noisy in gene expression and the processing and propagation of noisy signals within gene regulatory networks.

This chapter deals with an issue that gain much interest recently and that has been recognized of great important for the understanding of the dynamics of biological systems. Namely, a number of evidences have been found that stochastic mechanisms play a key role in the dynamics of biological systems. One such system is the genetic network. In gene networks, molecular-level fluctuations play a significant role. For example, they are a source of phenotypic diversity been individuals from a monoclonal population subject to the same environment. The degree of variability is highly influential on the population-level fitness.

In this chapter is also analyzed how cells receive noisy signals from the environments, detect and do their transduction with stochastic biochemistry. It is shown how mechanisms, such as cascades and feedback loops, permit the cell to manipulate noisy signals and maintain signal fidelity. Finally, from a biochemical implementation of Bayes's rule, it is shown how genetic networks can act as inference modules, inferring from intracellular conditions the likely state of the extracellular environment.

In the following Chapter 5, Zanudo and colleagues present a study of the dynamics of more abstract models of gene regulatory networks, namely Boolean Threshold networks. The Boolean network models have provide much insight into how gene regulatory networks dynamics is likely to dependent, on the large scale level, on features such as their topological and logical properties.

It is described how Boolean threshold networks have been useful tools to model the dynamics of genetic regulatory networks, including their applicability into accurately describe the cell cycles of S. cerevisiae and S. pombe. The dynamics of these models of gene networks is explored as a function of their mean connectivity, fraction of repression and activation interactions between genes and transcription factors, and various threshold values in the interactions between genes. One interesting aspect explored is how the value of threshold affects the dynamical regime of the network and the structure of its attractor landscape.

In Chapter 6, J. Kesseli and colleagues present a study on the ability of large scale gene regulatory networks models to exhibit complex behavioral patterns, as an emergent property of their ability to propagate information between the elements of the network. The stability and adaptability of gene networks in variable environments is investigated. Special emphasis is given to the role of information processing within the network as the source of the diverse, yet specific responses of the network. Such emergent macroscopic behaviors are governed by the interactions between the elements, thus, it is explored how the topology of the interactions determines the global dynamics of the system.

Unlike the traditional approaches, which define statistics regarding the structure or dynamics of complex systems, the authors propose a new approach derived from information theory so as to establish a unified framework to examine the structure and dynamics of complex systems of interacting elements.

In Chapter 7, F. Emmert-Streib discusses various local network-based measures in order to assess the performance of inference algorithms for estimating regulatory networks. These statistical measures represent domain specific knowledge and are for this reason better adapted to problems that are directly involving networks compared to other measures frequently used in this context like the F-score. He is discusses three such measures with special focus on the inference of regulatory networks from expression data. However, due to the fact that currently there is a vast interest in network-based approaches in systems biology the presented measures may be also of interest for the analysis of a different type of large-scale genomics data.

In Chapter 8, A. Ribeiro and J. Lloyd-Price investigate the information propagation within models of determinist and stochastic gene regulatory networks. It is known that cell types must be very restricted subsets of the possible states of a genetic network. As cells stably remain in such subsets of states, these are likely the dynamical attractors of the GRN. These attractors differ in which genes are active and in the amount of information propagating within the network.

From this perspective, the authors investigate what topologies maximize both the amount of information propagation between genes within noisy attractors, as well as the diversity of values of information propagation in each noisy attractor.

Measuring information propagation between genes by mutual information, we study in finite-sized Random Boolean Networks and in Delayed Stochastic gene networks how the dynamical regime of the network affects information propagation, quantified by pairwise mutual information between the temporal levels of genes' expression. It is shown how both the mean and the variance of values of mutual information in attractors depend on the dynamical regime of the network, and it is also investigated how the noise in the dynamics affects the propagation of information within the network.

It is argued that selection favors near-critical gene networks as these maximize mean and diversity of information propagation, and are most robust to noise. Phenotypic variation is critical since it is a necessary condition for evolution to occur. This principle is likely to apply not only to variability between organisms within a species, but also to the variability between cell types within an organism and the variability between cells within a cell type. Assuming that the fitness of gene networks depends on the ability to propagate information reliably between the genes, and that the fitness of organisms depends on cell-to-cell phenotypic diversity, the authors discuss whether near critical genetic networks are naturally favored.

In Chapter 9, S. Niiranen and colleagues first overview the state of art when it comes to the foundations of natural language and how it relates to the forms of communication used by other animals. Much work has been put into developing theories which would explain the structure of language and how it relates to the information processing capabilities of the human mind. Our ability to use natural language to communicate and co-operate with others is one of the defining characteristics of human intelligence. Second, they present a hierarchy of generic information processing systems and situate natural language in that hierarchy. Based on this they arrive at viewing, on a philosophical level, human language faculty as an embodied information processing system in the tradition of the embodied theory of mind.

3 Conclusion

In order to survive, organisms constantly survey their environment, being engaged in activities that can be perceived in terms of information processing or informational representation. From the simplest tasks, such as finding food, or avoiding a toxic substance so as to detect changes, information processing is essential for placing in practice any survival strategy. Thus, it is likely that biological systems for detection, managing and processing of information have been under constant selective pressure. How information is managed and processed by living organisms is still, to a great extent, a mystery. At best, we understand the functioning of some of the mechanisms responsible for the collection of information, and, at the lowest level, on how simple bits of information are processed and stored.

Within this book are described some of the biological systems and processes whose mechanisms of information processing and managing we best understand.

The goal of this book is to present an overview on the state-of-art on mechanisms of information processing in biological systems and to present the opinion of current leaders in the field on future directions of research. The primary audience of this book includes researchers and graduate students in computational biology, computer science, cognitive science, systems biology, life sciences in general, and medicine.

References

1. Godfrey-Smith, P., Sterelny, K.: Biological Information. Stanford Encyclopedia of Philosophy (1997)
2. Shannon, C.: A Mathematical Theory of Communication. Bell System Technical Journal 27, 379–423 (1948)
3. Godfrey-Smith, P.: Information in Biology. In: Hull, D., Ruse, M. (eds.) The Cambridge Companion of the Philosophy of Biology, pp. 103–119. Cambridge University Press, Cambridge (2007)
4. Schrödinger, E.: What is life? Cambridge University Press, Cambridge (1944)

Biological Information and Natural Selection

Daniel Cloud

Philosophy Department, 1879 Hall, Princeton University, Princeton, NJ 08544, USA

Abstract. What is 'biological information'? And how is it related to 'complexity'? Is the 'complexity' of a cell the same thing as the 'complexity' of a hurricane, or are there two kinds of complexity in nature, only one of which involves the transmission of 'information'? An account of biological information as a set of simplifying conventions used to coordinate the division of labor within cells is offered, and the question of why natural selection might tend to produce such systems of conventions is considered. The somewhat paradoxical role of mutations as occasionally-informative noise is considered, and a very simple formal model of the relationship between natural selection, complexity and information is developed and discussed in relation to adaptive landscapes. The question of whether we should expect evolution to slow down or speed up as the things that are evolving get more complicated is raised and dealt with. Finally, an epistemological moral is drawn, and briefly applied to the case of cancer.

In some ways, a cell is like a large, noisy, neural net.[1,2] Molecules, or segments of molecules genes, transcription factors, and the rest of the cell's regulatory apparatus act as nodes in the network. Interactions between molecules can be thought of as connections between the nodes. The propensity of a pair of molecules to interact can be identified with the weight of a connection. The pattern of response of a gene to the binding of a transcription factor seems analogous to the response function of a node. Either thing could in principle be modeled, somewhat imperfectly, with a huge, interconnected system of stochastic differential equations.

This mathematician's way of looking at cells a direct intellectual descendant of Stuart Kauffman's original 'random Boolean network' model of gene regulation [3] makes them seem rather like complex analog computers. And yet many biologists including Kauffman himself are skeptical of this further step. Are cells really just 'processing information'? Are computers really a good model for cells? Isn't life something rather different from computation? When we make this sort of idealized mathematical model, aren't we greatly oversimplifying what is, in reality, a very complex physical system? Is the information, as opposed to the molecules we mentally associate it with, really there at all?

On the other hand, though, if we *can't* legitimately talk about biological information, what sort of information is there left for us to talk about? We ourselves are living things, and all our information-processing devices are arguably just parts of our own extended phenotype. If we decide that it's meaningless or superfluous to speak of genomes as 'encoding information', or of gene-regulatory networks as 'processing information', then shouldn't the same arguments apply to neurons, and systems of neurons, and therefore to human thoughts, speech, and symbolic communication? Or is this taking skepticism too far, does this kind of biochemical nominalism throw the baby out with the bath water?

Some of these questions sound like philosophical ones. Perhaps that means that it's a waste of time to even think about them, but there really might be some utility in an attempt to at least state the obvious, to say what everyone already more or less knows, by now, about the nature of biological information. If this turns out to be easy to do, then we'll know that our ideas on the subject are in perfectly good order, and we can go back to doing science untroubled by any murky philosophical concerns. If it turns out to be difficult, we might end up having to do some thinking about philosophical problems, whether we want to or not. This seems like an elementary precaution, one that could hardly do us any harm.

1 A Criterion

Where do we start? If we want to know when it's legitimate to speak of encoded information in describing an event in nature, presumably we need to provide an account of what distinguishes cases in which it is legitimate to speak in that way from cases in which it is not. So one thing we are probably looking for is a criterion or rule that will allow us to decide which category a particular case belongs to.

A chromosome and a hurricane are both very complex things. But we tend to want to say that a DNA molecule encodes information in its sequence. On the other hand, we don't exactly ever want to say that a hurricane encodes much of anything. It's just a storm, it's random. Is there a principled distinction here, or merely an anthropomorphic habit of thought with respect to one sort of object but not the other?

There actually does seem to be a real difference behind this distinction. The information-encoding sequence of a DNA molecule, as opposed to the exact physical state of its constituent atoms at a particular instant, has a physically meaningful existence to the extent that the other machinery of the cell, RNA polymerases and transcription factors and things like that, would behave in the same way if presented with another molecule having that same sequence but whose individual atoms were in a slightly different exact physical state.

We can think of the sequence space associated with a biological macromolecule as a partition of the space of its exact physical states. Each sequence is associated with a very large number of slightly different exact physical states, all of which would be interpreted by the cell's machinery as representing the same sequence. It's meaningful to speak of this partition as something that really exists in nature because natural selection has optimized cells to act as if it does. Of course, if we had an omniscient, god's-eye view of the cell, we could still ignore the existence of genomic sequences as such and do a brute-force, atom-by-atom predictive calculation of its trajectory through state space, just as we could with the hurricane, but that's not what the cell itself is set up to do, that's not how it regulates itself. It uses much coarser categories.

Within certain limits, DNA molecules having the same sequence will be transcribed into RNA molecules having the same sequence, regardless of the exact physical state the parent DNA molecule is in. Of course, if it is folded in the wrong way, or methylated, or otherwise distorted or modified beyond those limits, a molecule may evoke a different response, but as long as it is within them, the same sequence will get the same transcription. Thus, it is the sequence of the parent molecule, and only the sequence of the parent molecule (not its various other physical features) that constrains the gross physical state of the daughter molecule. In that sense, it really is physically meaningful all by itself, because the rest of the machinery of the cell is optimized to treat genes with the same sequence as if they are the same thing. There just isn't anything like that going on in a hurricane, even though a cell and a hurricane are both 'complex systems'.

2 Two Types of Complexity

This certainly seems like an important difference. We can't really have much hope of answering our opening questions about how we should think of cells until we can explain it. Perhaps both things are complex systems, but, on an informal level, we habitually treat cells as being complex in a rather different way from hurricanes. For one thing, we never speak of optimization when talking about something like a hurricane. Optimization only comes in to a discussion when it somehow involves natural selection, not something that hurricanes are usually thought of as being subject to. Are there, then, two entirely different kinds of 'complexity' in nature, one naturally selected kind that involves 'information' and optimization, and another, non-self-replicating kind that involves neither? Why are they usually considered one phenomenon? Do we designate two entirely different things with one English word, 'complexity', or are the differences between the two types inessential?

Well, what could we actually mean by the word 'complexity', in the case of the hurricane? The general idea seems to be that because it's so big and contains so many different molecules moving in so many different directions at so many different speeds, it has quite a few degrees of freedom, that there are rather a lot of different physical states the system could be in. Turbulent flows can vastly magnify the effects of events that initially involve only a few molecules; so any and all of these degrees of freedom could affect the gross behavior of the system over time with indifferent probability. Predictions of the detailed behavior of the system far in the future will be inexact unless they are all taken into consideration. The absolute complexity of a hurricane might perhaps be equated with the volume of its statistical-mechanical phase space, a space with six dimensions for each molecule in the storm (three for position, and three for momentum along each of the three spatial axes) in which any instantaneous state of the whole system can be represented as a single point.

If that's what we mean by the word, though, we seem to have a bit of a problem. By that measure, any reasonably large collection of atoms is extremely complex, and cells are no more complex than drops of water of similar size. Both things are made of atoms, in approximately equal numbers, so the phase space for each should have a similar number of dimensions. A similar number of dimensions implies a roughly similar volume. Actually, though, we usually seem to want to treat cells as containing another layer of complexity beyond this obvious physical one, as being complex in a way that mere drops of water are not, as somehow having additional features. These additional features are precisely the kinds of things that seem to differentiate 'complex' entities like cells from 'complex' entities like hurricanes, so apparently it would be easier to keep our philosophical story straight if we were in a position to maintain that they actually exist in some physical way.

3 Simple and Complex Simplifications

The idea that these additional informational features must have some sort of genuine physical existence isn't exactly wrong, but we're going to have to adjust it a bit to make it coherent. From a physical point of view, all these 'additional' details cells supposedly have, though they're very complex as far as we're concerned, are actually the details of a set of ways in which they are simpler than oily drops of water. It's just that they're *simpler* in a very complicated way, unlike ice crystals, which are simpler than drops of liquid water in a very simple and uniform way. In both cases the actual physical entropy of the ordered object is lower than it would be if its constituent small molecules were all scrambled up and broken apart into a randomized liquid. Biological order, like classical crystalline order, is a simplification and regularization of a complex, irregular molecular chaos, one achieved, in the biological case, at the expense of creating even more disorder elsewhere. Repetition for example, the same kinds of amino acids and nucleotides repeatedly appearing everywhere in the cell and physical confinement (to compartments, or within the structures of macromolecules) are both conditions that, all other things being equal, decrease entropy. What increases, during evolution, is the complexity of the physical simplifications life imposes on itself and the world around it, not absolute physical complexity *per se*.

This rather obvious fact should help us make some sense of the idea, mentioned above, that biological information exists just when physically slightly different molecules with the same sequence are treated by the cell as being the same thing. To the extent that the machinery of a cell indifferently treats two different RNA molecules with the same sequence as if they were identical, it interacts with itself in a way that involves simplifying reality, *throwing away* some of the available details, responding in almost exactly the same way to many slightly different states of affairs. This is a necessary part of the process by which life continually re-imposes its own relative simplicity on the complex world around it. It's an energetically expensive imposition of sameness, an endothermic filtering-out of perturbations and suppression of irregularities. Some theoretical biologists call this general sort of process 'canalization', [4] though that term normally has a much narrower meaning.

Life responds to the members of the collection of possible states of affairs in the world in a way that is much simpler than that collection. In this sense, 'information', in a cell, isn't something that is *added* to the bare physical particulars of its constituent molecules, it's those bare particulars themselves with something expensively and repeatedly *subtracted* from them, a sort of topological invariant that can be conserved through all distortions and translations precisely because it is not very dependent on exact details. Biological information is filtered complexity, complexity that becomes information when some of its features are systematically and routinely discarded or suppressed as noise, when it is consistently responded to, by the cell's own machinery, in a way that only depends on some selected subset of its features.

(We might wonder, then, whether the idea of *transmission over a noisy channel*, so central to mathematical information theory, should really also be central to our theory of 'biological information', when *filtration* might seem like a better description of what's going on. Here, however, we may actually have succeeded in making a distinction without a real difference behind it, because the filtration of information is just the systematic ignoring or removal of noise, and that's what's required in transmission over a noisy channel as well. In the first few steps of visual perception, the conversion of light impinging on the eye to action potentials in neurons, the chaos of the external world is filtered down and transformed into a stereotyped, coded signal but this is also what a ribosome gets, via an mRNA, from the complex chaos of an actual, physical chromosome, a stereotyped, coded signal. There is no deep conceptual difference between the transmission of information over a noisy channel and its filtration.)

4   Codes as Conventions

The privileged subset of a complex molecule's features that matters typically consists of ones which have conventional mechanical relationships to other collections of features, elsewhere in the cell's machinery. A tRNA encodes a convention that relates a selected subset of the features of any mRNA to a selected subset of the features of some amino acid. In this way a codon 'represents' a particular type of amino acid. Only that type of amino acid will be allowed to randomly bounce into a situation where it will form a peptide bond with a growing protein chain, if that codon is currently the one being translated by the ribosome that is making the protein. The tRNA seems to encode a 'meaning'. Should this bother us, philosophically, in any way?

Yes and no. It shouldn't actually surprise us, but it should make us cautious. An analogy with a meaning in a human language does exist, but it's not a precise one. In this sort of discussion, we can't just pretend it isn't there, but on the other hand it can easily become misleading if we don't pay careful attention to the differences. That means that we need to dwell on it here, to make sure we understand its subtle pitfalls.

What, in general, is a 'meaning'? The philosopher David Lewis argued persuasively that we can often think of meanings, even in human languages, as conventions for solving what the game theorist Thomas Schelling dubbed 'coordination problems'.[5,6] Lewis argued that they can be thought of as public agreements committing us all, for the sake of convenience, to treat x's as valid proxies for y's under some circumstances, to treat the word 'cow' as a valid proxy for a cow, allowing us to collectively and individually reap the gains available from better coordination of our joint activities. Better-informed dairy farmers, who can speak to each other concerning cows and their affairs, mean that there is more butter for everyone than there would be if we couldn't discuss such things, and it is the traditional semantic convention that makes the noise a proxy for the animal that puts some of that butter on the table. tRNA molecules seem to function as 'traditional' semantic conventions in more or less this sense, since they facilitate the widespread use of particular codons as proxies for particular amino acids.

5 Selection, Coordination and Conventions

Why should natural selection, as a mechanism, tend to produce this sort of complex and rather artificial-seeming convention? What's already been said about Lewis's theory of meanings as conventions suggests a very general answer to the question. Somehow, to persist, grow, and reproduce, life must need to solve a lot of coordination problems, in a way not equally necessary to crystals and storms. But why is life, in particular, so beset with this kind of problem? Why don't crystals or hurricanes face a similar need for coordination across time and space, how is it that they can they grow and persist without transmitting coded information from point to point and time to time if cells can't?

That's a big question, but to make any progress in clarifying the analogies and dis-analogies between cells and computers, we must have an answer to it. To get one, we're going to have to back up a bit. Though it may seem like a digression, before we can really think clearly about how natural selection produces biological information and the conventional codes that carry it, we must first think a bit about the character of natural selection itself, as a process, and what sort of things it actually acts on.

Continual selective change in gene frequencies has to go on in a population over long periods of time for very much evolution by natural selection to occur. This can only really happen in a population that has both heritable variation and Malthusian dynamics. We tend to focus on heritable variation as the really essential thing about life. This way of thinking won't help us with our present inquiry, however, because the way inheritance works in actual modern organisms is already completely dependent on the existence of an elaborate form of encoded information. If we want to understand why living things create and process something recognizable to us as encoded information in the first place, we have to think carefully, instead, about the other side of Darwin's great idea, about the Malthusian dynamics of biological populations, and where those dynamics come from.

John Maynard Smith and Eörs Szathmáry have pointed out [7] that one feature of the molecular tool-kit on which life as we know it is based that seems to be essential is the fact that the basic building blocks amino and nucleic acids can be put together into an infinite variety of different shapes. This makes them a bit like a Lego kit, a set of standardized blocks with conventional shapes and affinities which allow them, collectively, to be assembled compactly into almost any form at all. On a molecular scale, 'almost any form at all' means about the same thing as 'catalyzing any reaction at all'. Because macromolecules made out of the same basic constituents can catalyze an enormous variety of different reactions, a self-replacing collection of such macro-molecules can divide the labor of self-replacement up into an arbitrarily large number of distinct reactive tasks, each of which will be performed far more efficiently, by the specialized catalyst that has evolved to take care of that aspect of the overall task, than any competing abiotic reaction.

In contrast, there's really only one kind of hurricane. The raw ingredients for a hurricane can only go together in one general way. The spiral can be larger or smaller, and can differ in various other insignificant ways, but the possibilities are very limited. A similar thing could be said of classical crystals at a given temperature and pressure, there tends to be only one form the crystalline lattice can take. Complex organic molecules are a much more promising substrate for evolution from this point of view, since a few types of small molecules can be assembled into an infinite variety of large ones.

Geometrically, this difference between classical crystals and large organic molecules might have something to do with the fact that carbon atoms are effectively tetrahedral, and therefore don't pack into three-dimensional space in any compact, periodic way unless the bond angles are distorted as they are in diamond. This fundamental geometric frustration, a three-dimensional analog of the five-fold symmetries that frustrate periodic tilings with Penrose tiles, is an intuitively appealing candidate for the role of culprit in life's complexity. Be that as it may, for our present purposes, it's enough to just register that the difference exists. This whole process of endlessly dividing up the self-replicative task faced by cells sounds very similar to something that Darwin talked about in the chapter on 'divergence of character' in *Origin of Species*.[8] There he describes the evolution of both ecosystems and systems of organs as an endless process of efficiency increasing through an ever-finer division of labor, which allows narrower and narrower tasks to be performed by more and more specialized structures. His arguments work just as well on the bio-molecular scale, inside of cells, as they did on the scale of whole ecosystems. The virtue of a flexible, Lego-like kit of tools which can evolve and conserve designs of arbitrary complexity is that it allows this division of self-replicative labor to go to its logical extreme.

The open-ended character of this project, its ability to accommodate contingencies with contingency plans and make lemons into lemonade, has made life a robust and common kind of self-organizing system, able to grow and persist in a wide variety of environments and survive a wide range of perturbations. The other kind of 'complex' (i.e., only slightly simplified) dissipative system in nature, hurricanes and crystals and things like that, depends on exactly the right conditions to grow, and will immediately begin to shrink if the environment deviates from those perfect conditions, but life can evolve to fit itself to almost any conditions it encounters. The difference is basically that between a toolkit that can only do one thing and a toolkit that can do anything, that can do many things, that can do several different interacting things at once, or several things one after the other.

Darwin's idea of an evolved, flexible division of labor is useful in a modern context because it lets us more precisely characterize the difference between the sort of self-replication a living thing engages in and the kind of 'self-replication' we see in crystallization, where the cellular structure of the crystalline lattice can create copies of itself very rapidly under just the right conditions. The molecules that make up biological systems replicate themselves by processes that are much more circuitous than those involved in the growth of crystals.

Circuitousness and a fine-grained division of labor are actually more or less the same thing. A division of synthetic labor means things are produced in many small steps, which is to say circuitously. Because catalysts can very efficiently accelerate many very specific reactions, this form of self-replication both is more powerful and more flexible than crystallization, better at grabbing atoms away from other competing geochemical processes under a wide variety of circumstances. It is this superior efficiency that makes it possible for living things to quickly fill an empty environment to the point of saturation. Thus, it is the flexibility of the underlying toolkit that ultimately explains the Malthusian dynamics of biological populations.

This circuitousness creates scope for alternate ways of doing things, and the possibility of selection between accidental variations. The more steps there are to a process, the more different ways there are to change a few of those steps, and the more different ways there are to change it, the more likely it becomes that one of those changes will improve its efficiency. Circuitousness also creates a need for coordination across time and space. Things must be done over here in time to get them ready for their role in a process going on over there, and if that process stops for any reason, they may have to stop as well. It is to these sorts of facts that the existence of anything recognizable as biological information must ultimately be attributed. The continual flow of information carried to the genome of a cell by transcription factors causes mRNA's to be produced when and where they are needed, and to stop being produced when they are no longer necessary. Circuitous self-replicators that have coordinated their activities most efficiently in time and space have had the most descendants, and in this way a whole genomic apparatus for coordinating and conserving circuitous patterns of molecular self-replication has evolved.

6 Mutations as Informative Noise

The division of the labor required for efficient homeostasis and self-replication requires coordination in the face of perturbation, and this requires the filtering out of noise. If cells were little robots made by tiny gnomes, our analysis could stop here. But life as we know it is the product of evolution. It would be a mistake to end our discussion of the relationship between noise and information in living things without considering the rather central case of mutations, a form of 'noise' which also seems to be the ultimate source of the genome's 'signal'. How can these two faces of the phenomenon be reconciled?

The existence of all the elaborate genome-repair adaptations found in modern cells shows that from one point of view that of individual cells mutations really are just noise, to be suppressed and filtered out if possible, like any other random noise. Of course, the cell does not actually have a 'point of view'. We should have said 'what typically maximizes the individual cell's fitness is the filtering out of as many mutations as possible.' Still, the fact that it has a fitness to maximize in the first place puts a cell much *closer* to having what we might normally think of as a 'point of view' than the other 'player' in the mutation/biological information game, Nature, in her role of selector of the fit and unfit.

We're sometimes tempted to personify this actor, too, at least on an unconscious level. Even Darwin used to talk this way occasionally, mostly for the deliberate Epicurean purpose of shocking us into seeing how different 'Nature the selector' really is from a mere person. Before he settled on the term 'natural selection', though, he considered the possibility of speaking, instead, in terms of 'the war of Nature'. ('Survival of the fittest' would then presumably have become something like 'continual repopulation by the victors', which at least has the virtue of descriptive accuracy.)

'Selection' may sound vaguely like a kind of information processing, but 'war' really does not. In fact, one of the most interesting things about natural selection is just exactly how *unlike* any normal form of 'information processing' it is. Still, somehow, out of this inchoate struggle a coded message eventually emerges. Our analysis of 'biological information' will be incomplete unless it includes some non-anthropomorphic explanation of how this magic trick works.

7 Serial Syntax Meets Holistic Semantics

Much of the way we habitually think about information and information processing comes from the fact that we ourselves are living things, and as such, are not infinitely capable. We have to break computational tasks down into pieces, and deal with one piece at a time. Bandwidth is a real constraint in the informational dealings of mere creatures, and everything they do is arranged to get around that fact. It takes us time to read a book, just as it takes an RNA polymerase time to produce an error-free transcript of a gene. There are many small steps involved. Whenever we want to do anything complicated, we have to do it in small chunks. Messages must be arranged in a sequential way, one simple piece after another, so that that they can be transmitted and decoded a little bit at a time.

When we think about computational problems, we of course assume that it will take a certain amount of time or space to solve a problem of a given complexity, and we are very interested in classifying such problems with respect to this kind of difficulty. But nature, as selector, does not have to respect these classifications, because it is not an information-processing device solving problems of limited complexity sequentially in stages. The 'war of Nature' happens in the real world, so the whole idea of bandwidth is irrelevant. A predator may kill its prey in an arbitrarily complicated way. This doesn't make it any harder, or any more time consuming, for the prey to die. The prey doesn't have to use any computational resources, or do any work, to be affected by the predator's arbitrarily complex strategy, and the event doesn't have to take any particular amount of time. Indirect ecological interactions acorns from masting oaks feeding mice who bring ticks with Lyme disease to deer of arbitrary complexity can also affect fitness in serious ways. There is no increase in computational cost associated with passing through more complex versions of this sort of filter.

Using terms like 'select' and 'evaluate' conveys a somewhat misleading impression of the way natural selection works. Natural selection is not a cognitive or computational process at all. We face a temptation to imagine it as having some of the limitations of such a process, but it does not. Evolution has not made cells what they are today by breaking them down into pieces and evaluating or dealing with the pieces separately, one a time. Recombination does break repeatedly genomes down into pieces of variable size, but 'evaluation' selection itself mostly happens to entire cells and organisms, and to genes in the context of whole cells and organisms, embedded in whole, particular possible worlds. It takes a whole organism, in all its complexity, in a whole complex world, to live, or die, or raise a litter of kittens. (Will the rains fail this year?) Genghis Khan has a large number of descendants today because of who he was in all his individual complexity, and what 13th century Central Asia was like. His Y chromosome might not have done as well in modern Chicago.

Natural selection is not the kind of mechanism that evaluates things a little bit at a time, it is the sort of mechanism that evaluates them all at once. The information in cells is processed serially, but it is first produced by natural selection as a gestalt, a single tangled-up whole. (It normally takes many selection events to bring a gene to fixation in a population, but each of those events involves a whole individual's complex struggle to get through a whole life in a whole world with a whole genome, so this does not mean that the process has somehow been 'broken down into smaller pieces' in the relevant sense.)

Whether or not a mutation is new biological information or just more noise isn't even something that can generally be discovered just by inspecting the organism in which it occurs. What is adaptive depends on the environment that an organism finds itself in; the same alteration to the same genome can be fatal or providential depending on external circumstances. The parts of the genome coding for Darwinian preadaptations are noise until the environment changes in a way that makes them useful, at which point they become signal. This determination may depend on any feature of the environment, even one as tiny as a virus. Its full complexity is constantly available as a filter to turn mutational noise into biological information, and neither of these categories is particularly meaningful in the absence of that full complexity.

Biological information only counts as biological information when considered as part of an entire detailed world. This makes it look, to a philosopher, something like 'truth', a property sentences only have at particular possible worlds. Because we process the syntactic aspects of language serially, and because computers can be built that do the same thing in the same way, we can easily forget that semantic interpretations are holistic in character, in that they require whole worlds or collections of worlds in the background for sentences to be true or false at. But in logic as we now understand it, there is no just such thing as a semantic interpretation without a 'model' standing behind it. Sentences are only true or false in the context of whole worlds or whole theories, and similarly biological information is only really distinct from meaningless noise in the context of a particular environment in all its fine grainy detail, down to the last pathogen.

8 An Oracle for Solving Decision Problems

Is there a more formal way of thinking of the process of natural selection? If we did want to try to describe it in the language of information processing, how would we do it? We can actually do it fairly well with a model of computation that is ridiculously strong, one usually considered too strong to represent any actual information-processing system, a model which tends to be presented as a limiting case showing what could happen if certain necessary assumptions were relaxed. From a formal point of view, natural selection is something like an oracle for solving a set of decision problems.

A decision problem is any logical problem that can be posed in a form that admits a yes-or-no answer. An example is Goldbach's conjecture, the assertion that every even number greater than two is the sum of two primes. The conjecture is either true, or false; which it is is a decision problem. An organism either contributes its genes to the next generation or it doesn't which makes fitness a yes-or-no proposition, or anyway lets us treat it that way as a first approximation.

An oracle is an imaginary computational device that, presented with a putative answer to any computational problem, can infallibly evaluate it as correct or incorrect in a single computational step. An oracle that could infallibly tell whether a yes-or-no answer to any decision problem was the correct one could resolve thorny mathematical questions like Goldbach's Conjecture in no time at all. We would just guess 'yes, every even number greater than two is the sum of two primes', and the oracle would tell us if we were right or wrong. Of course, we still wouldn't know why we were right, but evolution doesn't explain itself either. We have to reverse-engineer our way to the whys of things, as we would if an oracle told us that Goldbach's Conjecture was certainly true. We can think of the organism as the putative answer to the problem of how to deal with its environment, and successful reproduction as an evaluation as correct. There is no minimum number of steps this process must involve, and no dependence of the time required on the complexity of the decision problem being solved, so only an oracle will do as a formal model.

9 Oracles and Adaptive Landscapes

To tease out the full implications of this simple way of conceiving of natural selection, we need to think about a problem that came up in the 1960's about a particular type of evolution. At that time, some doubt was expressed that anything as complicated as a protein molecule could possibly have evolved in the amount of time available.

With twenty possible amino acids at each locus, a relatively modest chain of 100 amino acids has about 10^{130} possible sequences. Since the universe is only about 10^{18} seconds old, it seems as if an unreasonably large number of sequences would have to be tried out each second to find a particular sequence with a particular function in the amount of time available.

John Maynard Smith demonstrated, however, that this supposed difficulty is more imaginary than real. [9] He did it by making an analogy with a simple word game. Suppose we're playing a game in which we start with some four-letter English word Maynard Smith illustrated his argument by starting with 'word' itself. The objective is to transform that word into some other actual English word through a chain of intermediates that are themselves all valid words. His example was 'word wore gore gone gene', but of course there are plenty of other examples we could give, one being 'love lave have hate'.

There are 26 letters in the alphabet, so the space of four letter English words contains 264 or 456,976 points. However, no point is more than four changed letters away from any other. Whether or not it is possible to get from point a to point b in this space in a given amount of time is not a matter of how many points there are in the space in total, it is a matter of whether or not there is a valid bridge between the two words. Whether the space as a whole is easy to get around in depends on whether there is a percolating network of such bridges all through the space, which will take you from almost any point to almost any other in a few steps. There may be other valid destinations also reachable from the starting point by chains of valid intermediates, and there may be dead ends, and detours, all of which complicates the statistics a bit, but there is certainly never going to be any need to try out half a million words just to get from one point to another.

Why is the problem so much simpler than it initially seemed? It's easy to mistake this for a counting argument of some kind, to suppose that the person who was raising difficulties about the time it would take to search these sorts of combinatoric spaces was simply making a mistake about the numbers. But that isn't really what's going on here at all. What makes the word game easier than it otherwise might be is the fact that we all speak English, and therefore are able to instantly recognize that some of the four-letter combinations adjacent to a given word are themselves valid English words, while others are not. We aren't really blundering through the space at random. Each successive step involves a process in which we evaluate the available candidates to determine whether they are actually words. So the evolving word has to pass repeatedly through the filter of our brain, and pass a test for validity on the basis of knowledge stored there. Paths that terminate in dead ends don't get searched any further.

We don't need to sample every point in the space, we just need to search down a tree.

Here, our knowledge of what counts as a valid English word is playing the same role that the environment plays in the process of natural selection. Both act as filters, weeding out unsuitable candidates. The word itself doesn't do any 'information processing' in this process. It is subjected to the decisions of an external oracle, which gives a yes-or-no answer to the question of whether or not a proposed next step is actually a meaningful English word in its own right. It is in this sense that this sort of process can be thought of as the solving of a series of decision problems by an oracle. Similarly, the evolving protein itself doesn't need to 'know' whether an adjacent sequence represents an adaptive variant of the original design, or what it might be useful for. The environment will supply that information instantaneously and for free, by either killing a mutant cell or letting it prosper.

To appreciate the enormous creative power of natural selection, we really only need to ask one further question. What happens to the process of evolution, in this sort of combinatoric space, if the evolving objects get more complicated?

An obvious and intuitive way to extend the analogy would be to move from thinking about the space of four letter English words to the space of eight letter English words. As the examples of the reachability of 'love' from 'hate' and of 'gene' from 'word' illustrate, it is relatively easy to get from word to word in the space of four-letter words through a continuous chain of viable intermediates. However, there are also isolated islands like ALSO and ALTO, a pair of words that is not connected to any other word. Do these sorts of isolated islands become more or less common as the number of letters in each word goes up?

Take any eight-letter word say, CONSISTS or CONTRARY. Does it have a meaningful single-mutation neighbor? Typically, no. It is usually impossible to move from eight-letter English word to eight-letter English word through a continuous chain of valid intermediates. So it might easily seem to us as if this sort of evolution ought to get more difficult, ought to slow down, as the evolving objects get more complicated. But this is precisely where the analogy between natural selection and any kind of cognition breaks down most badly.

The greater difficulty in navigating through the space of eight-letter words is not a consequence of some innate topological property of high-dimensional spaces. It is a consequence of the limits of the human mind, and the consequent simplicity of human languages. Natural human languages have, at most, on the order of 10^5 words. The space of eight-letter sequences contains on the order of 10^{10} sequences. Naturally, since less than $1/10^5$ of the sequences in the space are meaningful words, and since each word only has 208 neighbors, most meaningful eight-letter English words will not have even a single meaningful neighbor. But the evolution of proteins is completely different. Nature just doesn't face this sort of cognitive limitation. It's not as if she can only remember the designs of a few thousand proteins. She doesn't remember things at all, that's not how this particular oracle does its magic. Anything that actually happens to work in competition, that in some way, however improbable-seeming, leads to the production of adult offspring, simply works.

In the absence of human cognitive limitations, what actually happens to this sort of space as it acquires more dimensions? There is no a priori reason why viable protein designs must be vastly more sparsely distributed in a space of longer sequences than they are in a space of shorter ones. The change in their density depends on things like the degree of modularity in the molecule's design, and the commonness of neutral mutations. Suppose, as a limiting case, that there is a constant, unvarying probability per locus, p, that some mutation at that locus leads to an adjacent sequence which is as adaptive as or more adaptive than the original sequence. As the length of the evolving sequence increases, there are more loci on it. All other things being equal, the probability that *some one* of these loci can be mutated in some adaptive way should go up as the number of loci increases. Call the length of the evolving protein, the number of loci, therefore the dimensionality of the sequence space, L. The probability that there is some adjacent sequence that can be moved to without loss of fitness is then just Lp. This quantity goes up linearly with the length of the evolving chain. At some point, it reaches 1.

As it becomes more likely that, for any given sequence in the space, *some one* of the adjacent sequences is as fit or fitter, it also becomes much more likely that the space as a whole is permeated by a percolating network of adjacent sequences each of which is just as fit as its neighbor. Thus, as the length of the evolving chain goes up, the space becomes easier and easier to get around in, not harder, as intuitions derived from the way we humans filter information might suggest.

On the basis of this model, it seems possible to argue that there may actually be a *minimum* complexity below which things cannot evolve efficiently. There may be countervailing effects that in fact cause p to fall in most cases as L goes up, but if the decline is less than proportional to the increase in L, percolation still becomes more probable as L goes up. No doubt some times it is, and some times it isn't; the sorts of evolved complexity we actually see in the world around us should be biased in favor of those cases in which p declines at a less than proportional rate.

How fast p declines with rising L depends on the precise characteristics of the thing that is evolving. If mutations typically have very small effects on phenotype, p should be fairly large no matter how complex the evolving object is. Modular designs limit the extent to which p depends on L. The modularity of life on a molecular scale may have more to do with limiting the decline of p than anything else. Neutral percolation can continue to be easy no matter what L is, if the evolving molecules are set up in a way that means that many mutations have no great effect on fitness. None of this seems to put a dent in the counterintuitive conclusion that evolution should actually often get easier, not harder, as the evolving object becomes more complicated. Adding degrees of freedom makes discrete combinatoric spaces more densely interconnected, not less, so in the absence of human cognitive limitations or other countervailing effects, free evolutionary percolation should become more and more likely as the evolving entities get more complex.

This conclusion, while it seems to be inescapably implied by Maynard Smith's model, may strike some readers as repugnant to common sense. Perhaps it is, but on the other hand, it does seem to fit fairly well with the facts. As far as we can tell, the evolution of life on Earth is not particularly slowing down as the evolving organisms and ecosystems get more complicated. There has been rather a lot of complex evolution since the Precambrian, or in other words during the last ten percent of life's history on earth. Things seem to have moved much more slowly when the evolving organisms were simpler. The time needed to evolve a human's brain from a lemur's was actually much shorter than the time needed to evolve a lemur's brain from that of a fish.

A case could even be made that evolution has speeded up significantly as the evolving organisms have become more complex. An awful lot has happened to life since the relatively recent event of the rise of the flowering plants. The whole evolutionary history of the apes and hominids has occurred in the last few percent of the Earth's history. The apparent implications of Maynard Smith's model and the data from the fossil record are actually in complete agreement. What the model seems to predict is what we actually observe. It's common sense, derived from our own human experience of designing things using our own limited cognitive capabilities, that makes the oracular power of the selecting environment, its ability to turn any problem no matter how complex into a decision problem which it can solve in no time at all, seem so unlikely.

Things get harder and harder for us to redesign as they get more complicated, because we're a little stupid. But there's no upper limit on the complexity of the systems natural selection can optimize, and no necessary dearth of extremely complex structures and ways of doing things in nature's infinite library of random designs. Complexity, as Kauffman first pointed out many years ago, [10] is available to nature for free. All the information represented by the human genome was obtained 'for free' from an oracle but actually using it, and sending it from place to place in the cell, is energetically very expensive.

10 An Epistemological Moral

What use is it to know all this? Does all this philosophical reflection, all of this very explicit and careful restating of the obvious, actually help us in any way in, say, understanding cancer? In fact, there's an epistemological moral here for the sciences of complexity in general, one that is in fact particularly salient to the way we think about cancer. The moral of the story is that biological complexity is not at all simple, that it's actually really, really complicated.

The way we've done physics, since the time of Descartes and Newton, is by assuming that behind the apparent complexity of nature, there's actually a deep simplicity, that if we can read a set of rules off of the behavior of some physical system we're likely to have found a complete specification of its nature that is much simpler than the system seemed at first. We might think of this approach, which has worked so remarkably well in physics, as reflecting a sort of unconscious commitment to Platonism on the part of physicists, since the essence of Platonism is the idea that the apparent variety, changeability, and complexity of the world we experience is just the result of varying combinations of much simpler and more fundamental base-level entities which can not be perceived with the senses. On the basis of this commitment, we really have been able to discover very simple rules we call them 'fundamental physical laws of nature' which can be stated, in their entirety, in mathematical language, and which actually do completely determine the character of the systems they govern. One electron can be treated, by us, as being pretty much the same as another because they really are all pretty much exactly the same and will both behave in exactly the same way in exactly the same situation. The most accurate description of nature and the simplest one often coincide in physics.

Because so much of the apparent complexity of nature has revealed itself to be the product of these sorts of simple physical laws, we now have the expectation, as scientists, that behind complexity in general there is always an underlying simplicity, and that once the symbolically-expressible rules governing a system are read off of its behavior, we will know all about it. To think this way about cells, however, is to make too much of the analogy between the sort of rules that 'govern' the behavior of electrons and the sort of rules that 'govern' the expression of a gene.

Behind the apparent complexity of the electron's behavior is a simple set of rules. Once we have them, we know all about it. The expression of a certain gene may also seem complex, and yet we may discover that it, too, seems to follow certain simple rules. But, though it may well be true that behind the apparent complexity of the gene's behavior there is some set of simple underlying regularities, it's also certainly going to be true that behind these apparently simple underlying regularities there is actually even more physical complexity lying in wait. The simple conventional rules that govern the behavior of biologically meaningful categories of molecules in cells are not the ultimate physical 'foundations' on which the cell's dynamics are based, they are the complex biomechanical consequences of everything else going on in the cell.

The idea that a certain sequence is reliably transcribed in a certain way can only be pushed so far; at some point, physical differences in the molecule whose sequence it is say, its folding state start to matter, and a qualification must be added to the rule. But that qualification will not be a simple one; it must consist of an account of all the various things that can physically happen in the cell when the molecule is in all the various abnormal states that are physically possible for it. This is why the cellular mechanics of cancer are so difficult to pin down; once the cell's machinery is no longer functioning in the usual way, there are an unimaginably vast number of other things it could be doing. As it turns out, lots and lots of them are ways of being cancer.

In this huge wonderland, anything at all about the cell's state could end up mattering in almost any way. We are outside the narrow conventions of the normal human cell in cancer, and the volume of parameter space that lies available for exploitation just outside those boundaries is huge. *We* immediately get lost in these vast new spaces, but the oracle of natural selection already 'knows' all the paths through their complexities without actually needing to know anything at all. Since natural selection acts on the whole phenotype of the cell all at once, there is no particular reason why the adaptive effects of a genomic abnormality should be simple enough for us to easily understand. Every kind of cancer is its own vast universe of complexity, and each and every single tumor may end up being incomprehensibly complex. There is no upper bound on the complexity of the new adaptations that can be picked out and amplified by this ignorant but unerring process, because there are no cognitive limitations to create such a bound.

Probably the most surprising thing about metastasis is that it can evolve at all, that such a high fraction of the cancers that naturally occur in human beings are able to develop elaborate adaptations for spreading through the body and thriving in new locations over periods of a few months or years. Of course, already having a full human genome, cancer cells have a big head start on the project of evolving new adaptations for thriving inside the human body. But what must make the evolution of metastasis possible is the overlay of a huge percolating network of equally viable, perhaps only neutrally different cellular phenotypes on human genome space. As creatures, with limited capacity for processing information, we seem to vastly underestimate the density of new, workable complex designs in the realm of unrealized forms, supposing that there are only a few, and fewer, proportionally, as complexity increases.

(In a habit that goes back millennia, we suppose that the things that *are* must be some reasonably large fraction of the things that *could be*, though in fact they are a vanishingly small and rapidly shrinking sub-set. Evolution explores a combinatoric space that gets easier to move around in in a way that depends more or less linearly on its dimensionality, while the number of points in that same space increases exponentially with its dimensionality, so an evolving life-form, as it becomes more complex, moves more and more freely through a vaster and vaster set of possibilities, exploring smaller and smaller fractions of them at greater and greater speeds.)

Because of this inappropriate, lingering Platonic assumption (in most kinds of Platonism, the Platonic forms are supposed to be fewer than the things in the world, though in *Parmenides* Plato admits that this supposition is problematic) we look for simple explanations of the adaptive effects of abnormalities, and we tend to think there must only be a few kinds of colon or lung cancer. But this is an illegitimate transfer of the sort of intuition we derive from our own experiences with things like eight-letter English words whose profusion is limited by our stupidity onto a universe that is not stupid (or clever) at all.

The rules we can deduce about how a cell will respond to a particular perturbation will only be simple if we are willing to allow a multitude of un-described possible exceptions. Any attempt to describe *all* of the possible exceptions to such a rule would involve describing all of a cell's possible physical states in exact detail. Then we would be no better off than we are in dealing with a hurricane. Exact prediction of future events is only possible when the simplifying conventions work as they're supposed to. The rules that govern cells are only simple when they are imperfectly stated; anything simple we say about them is only a sort of executive summary, an approximate characterization of what's 'usual' for a system whose real behavior can become arbitrarily different in particular cases. No matter what things are like in physics, the simplest description of a system and the most accurate one do *not* ever coincide in biology. A fully accurate description of a biological system is a description of that system in all its unique complexity. (Of course, such a description is impossible but that just means biologists will never run out of things to do.)

This makes many of the regularities we discover in the living world very different from the sorts of laws studied by physicists, and the kind of knowledge we can have about living things very different from the kind of knowledge we can have of the fundamental physical laws. Those things are actual necessities of nature, descriptions of universal mechanisms that couldn't possibly work in any other way. The regularities we observe in cells really are just local conventions, which serve to allow coordination between large numbers of molecules, when they actually obtain, which is not all the time, or in all cells. In death, all these regularities decay and disappear. As far as we know, there is never any such breakdown of the rules governing electrons.

So a large, noisy neural net might in fact be a useful model of information processing in cells, but only if we keep in mind the caveat that the system will act like one all the time except when it doesn't, when the underlying mechanical properties of the components come to matter in ways a neural net model doesn't reflect. It's very optimistic to suppose that we can ever arrive at a detailed simulation or model of cellular function that also covers all abnormal cases, that accurately predicts the system's response to any perturbation, because anything at all can matter in an abnormal case, and the suppression of idiosyncratic detail is the essence of simulation. The only fully accurate model of a cell as information processor is the cell itself in all its complexity, in its natural environment, where anything at all can happen, and information as such does not really exist.

References

1. Ben-Hur, A., Siegelmann, H.: Computing with Gene Networks. Chaos (2004)
2. Siegelmann, H.: Neural Networks and Analog Computation: Beyond the Turing Limit. Birkhauser, Boston (1998)
3. Kauffman, S.: Metabolic stability and epigenesis in randomly constructed genetic nets. Journal of Theoretical Biology 22, 437–467 (1969)
4. Kauffman, S.: The large scale structure and dynamics of gene control circuits: an ensemble approach. Journal of Theoretical Biology 44, 167–190 (1974)
5. Lewis, D.: Convention: a Philosophical Study. Blackwell, London (2002)
6. Schelling, T.: The Strategy of Conflict. Harvard Press, Boston (1960)

7. Maynard-Smith, J., Sathmáry, E.: The Major Transitions in Evolution. Oxford University Press, Oxford (1995)
8. Darwin, C.: Origin of Species. Charles Murray, London (1859)
9. Maynard-Smith, J.: Natural selection and the concept of a protein space. Nature 225, 563–564 (1970)
10. Kauffman, S.: The Origins of Order. Oxford University Press, Oxford (1993)

Swarm-Based Simulations for Immunobiology
What Can Agent-Based Models Teach Us about the Immune System?

Christian Jacob[1,2], Vladimir Sarpe[1], Carey Gingras[1], and Rolando Pajon Feyt[3]

[1] Dept. of Computer Science, University of Calgary, Canada
[2] Dept. of Biochemistry and Molecular Biology, University of Calgary, Canada
[3] Center for Immunobiology and Vaccine Development, Children's Hospital Oakland Research Institute, Oakland, CA, USA

Abstract. In this contribution, we present a computer model of information processing within a highly distributed biological system of the human body, which is orchestrated over multiple scales of time and space: the immune system. We consider the human body and its environment as a well-orchestrated system of interacting swarms: swarms of cells, swarms of messenger molecules, swarms of bacteria, and swarms of viruses. Utilizing swarm intelligence techniques, we present three virtual simulations and experiments to explore key aspects of the human immune system. Immune system cells and related entities (viruses, bacteria, cytokines) are represented as virtual agents inside 3-dimensional, decentralized compartments that represent primary and secondary lymphoid organs as well as vascular and lymphatic vessels. Specific immune system responses emerge as by-products from collective interactions among the involved simulated 'agents' and their environment. We demonstrate simulation results for clonal selection in combination with primary and secondary collective responses after viral infection. We also model, simulate, and visualize key response patterns encountered during bacterial infection. As a third model we consider the complement system, for which we present initial simulation results. We consider these *in-silico* experiments and their associated modeling environments as an essential step towards hierarchical whole-body simulations of the immune system, both for educational and research purposes.

1 In Perspective: Modeling with Bio-agents

Major advances in systems biology will increasingly be enabled by the utilization of computers as an integral research tool, leading to new interdisciplinary fields within bioinformatics, computational biology, and biological computing. Innovations in agent-based modelling, computer graphics and specialized visualization technology, such as the CAVE© Automated Virtual Environment, provide biologists with unprecedented tools for research in virtual laboratories [9,20,38].

Still, current models of cellular and biomolecular systems have major shortcomings regarding their usability for biological and medical research. Most models do not explicitly take into account that the measurable and observable dynamics of cellular/biomolecular systems result from the interaction of a (usually large) number of 'agents'. In the case of the immune system, such agents could be cytokines, antibodies, lymphocites, or macrophages. With our agent-based models [27,50], simulations and visualizations we introduce swarm intelligence algorithms [6,11] into biomolecular and cellular systems. We develop highly visual, adaptive and user-friendly innovative research tools. Our hope is that these agent models will gain broader acceptance within the biological and life sciences research community — complementing most of the current, more abstract and computationally more challenging[1] mathematical and computational models [43,7]. We propose a model of the human immune system, as a highly sophisticated network of orchestrated interactions, based on relatively simple rules for each type of immune system agent. Giving these agents the freedom to interact within a confined, 3-dimensional space results in emergent behaviour patterns that resemble the cascades and feedback loops of immune system reactions.

With the examples presented here, we hope to demonstrate that computer-based tools and virtual simulations are changing the way of biological research. Immunology is no exception here. Nowadays, computers are becoming more and more capable of running large-scale models of complex biological systems. Recent advancements in grid computing technologies make high-performance computer resources readily accessible to almost everybody [40]. Consequently, even highly sophisticated — and to a large extent still poorly understood — processes, such as the inner workings of immune system defense mechanisms, can now be tackled by agent-based models in combination with interactive visualization components. These agent models serve as an essential complement to modeling approaches that are traditionally more abstract and purely mathematical [43,7]. Making simulation tools (almost) seamless to use for researchers and introducing such tools into classrooms in bioinformatics, systems biology, biological sciences, health sciences and medicine greatly increases the understanding of how useful computer-based simulations can be. Such models help to explore and facilitate answers to research questions and, as a side effect, one gains an appreciation of emergent effects resulting from orchestrated interactions of 'bio-agents'.

In this contribution, we present three examples of our swarm-based simulations, which, we think, fulfill these criteria. With these demonstrations, we have implemented an interactive virtual laboratory for exploring the interplay of human immune system agents and their resulting overall response patterns. The remainder of this chapter is organized as follows. In Section 2 we introduce the immune system from a biological perspective as a highly distributed architecture of networked structures and associated bio-agents. In Section 3 we give an overview of related simulation and modeling approaches regarding

---

[1] For example, many differential equation models of biological systems, such as gene regulatory networks, are very sensitive to initial conditions, result in a large number of equations, and usually require control parameters that have no direct correspondence to measurable quantities within biological systems [7].

immune system processes. In section 4, we discuss a first version of our agent- or swarm-based implementation of the immune system, highlighting the modelled processes and structures. Section 5 gives a step-by-step description of both simulated humoral and cell-mediated immunity in response to a viral antigen. Memory response, which we analyze in more detail in Section 6, shows the validity of this first model showing reactions to a second exposure to a virus. In Section 7 we reflect on this first agent-based model, discussing its validity, advantages and short-comings. In the context of a second, and expanded model, we give an information and biological perspective of the decentralized immune defenses in Section 8. The key design aspects and main results of our simulation system — now with an expanded model — are described in Section 9. Here, we discuss a second set of simulation experiments for clonal selection after primary and secondary exposure to viral infections. As an extension to this second model, we show simulations of immune system reactions to bacterial infection. A review of lessons learnt from this expanded model is presented in Section 10. In Section 11 we investigate a swarm model of the complement system, which, from an evolutionary perspective, displays the most ancient form of immunity. The complement system also represents a bridge to adaptive immunity. In Section 12, we conclude this contribution with a summary of our work and suggestions for the necessary next steps towards an encompassing simulation environment to study, investigate, and explore immune system processes.

2   The Immune System: A Biological Perspective

The human body must defend itself against a myriad of intruders. These intruders include potentially dangerous viruses, bacteria, and other pathogens it encounters in the air and in food and water. Our body also has to deal with abnormal cells that are capable of developing into cancer cells. In reaction to these potential threats to the body's survival, Nature has evolved two cooperative defense systems:

- **Nonspecific Defense:** The nonspecific defense mechanism does not distinguish one infectious agent from another. This nonspecific system includes two lines of defense which an invader encounters in sequence. The first line of defense is external and is comprised of epithelial tissues that cover and line our bodies (e.g., skin and mucous membranes) and their respective secretions. The second line of nonspecific defense is internal and is triggered by chemical signals. Antimicrobial proteins and phagocytic cells act as effector molecules that indiscriminately attack any invader penetrating the body's outer barrier. Inflammation is a symptom which results from deployment of this second line of defense.
- **Specific Defense:** The specific defense mechanism is better known as the *immune system* (IS), and is the key subject of our simulations. The immune system represents the body's third line of defense against intruders and comes into play simultaneously with the second line of nonspecific defenses. Defense mechanisms from the immune system respond specifically to a particular

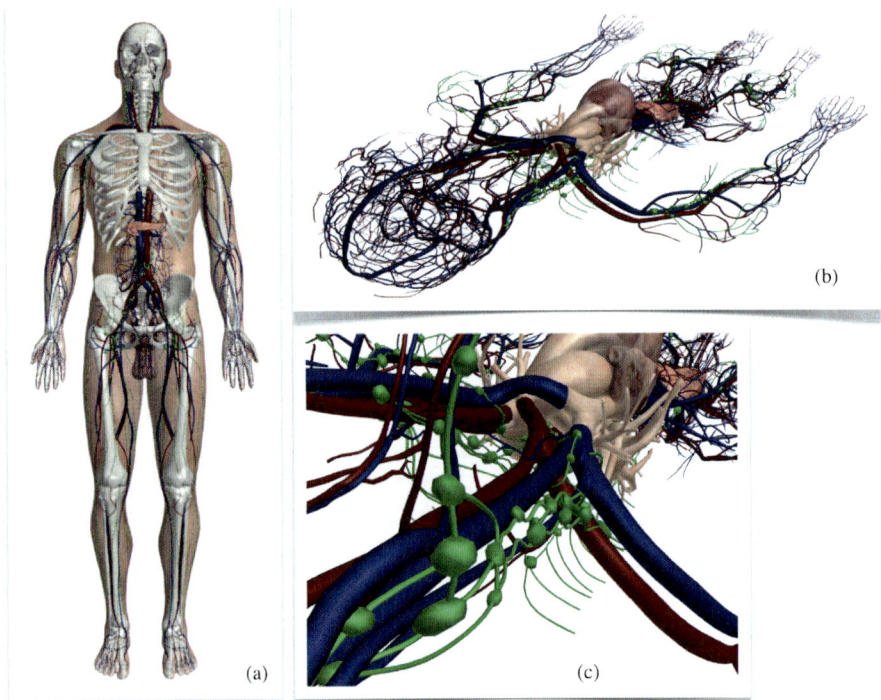

Fig. 1. The immune system as an anatomical and physiological network within the human body: (a) Full-body view with lymphatic system, circulatory system, skeletal system, spleen, thymus, and lymph nodes; (b) close-up view of the circulatory and lymphatic system; (c) illustrating lymph nodes and blood vessels in the right neck area. These anatomical models are rendered using Lindsay Presenter (http://lindsayvirtualhuman.org).

type of invader. For example, an immune response includes the production of antibodies as specific defensive proteins. Such as response also involves the participation of white blood cell derivatives (lymphocytes). While invaders are attacked by the inflammatory response through antimicrobial agents, and phagocytes, they inevitably come into contact with cells of the immune system. These IS cells, in turn, mount defenses against specific invaders by developing a particular response against each type of foreign substance, such as microbes, toxins, or transplanted tissue.

2.1 Humoral and Cell-Mediated Immunity

The immune system mounts two different types of responses to antigens: a humoral response and a cell-mediated response (Fig. 2). *Humoral immunity* results in the production of antibodies through plasma cells. The antibodies circulate as soluble proteins in blood plasma and lymph. The circulating antibodies of the humoral response defend mainly against toxins, free bacteria, and viruses

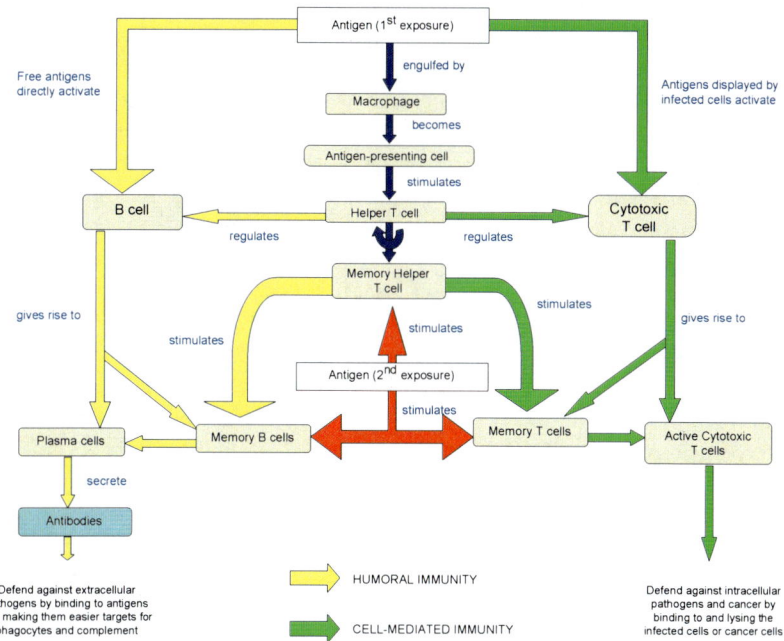

Fig. 2. Schematic summary of immune system agents and their interactions in response to a first and second antigen exposure. The humoral and cell-mediated immunity interaction networks are shown on the left and right, respectively. Both immune responses are mostly mediated and regulated by macrophages and helper T cells (reproduced from [23]).

present in body fluids. *Cell-mediated immunity* depends upon the direct action of certain types of lymphocytes rather than antibodies. In contrast, lymphocytes of the cell-mediated response are active against bacteria and viruses inside the host's cells. Cell-mediated immunity is also involved in attacks on transplanted tissue and cancer cells, both of which are perceived as foreign.

2.2 Cells of the Immune System

There are two main classes of white blood cells (lymphocytes), which are key players in the adaptive immune system response processes: *B cells* are involved in the humoral immune response; *T cells* are involved in the cell-mediated immune response. Lymphocytes, like all blood cells, originate from pluripotent stem cells in the bone marrow. Initially, all lymphocytes are alike, but eventually they differentiate into T or B cells. Lymphocytes that mature in the bone marrow become B cells, while those that migrate to the thymus develop into T cells. Mature B and T cells are concentrated in the lymph nodes, spleen and other lymphatic organs. It is within these specialized organs, that act as concentrating hubs, where the lymphocytes are most likely to encounter antigens.

Both B and T cells are equipped with antigen receptors on their plasma membranes. When an antigen binds to a receptor on the surface of a lymphocyte, the lymphocyte is considered activated and begins to divide and differentiate. This gives rise to *effector cells*, those cells that actually defend the body in an immune response. With respect to the humoral response, B cells activated by antigen binding give rise to *plasma cells* that secrete *antibodies*. Antibodies are very specific and help to eliminate a particular antigen (Fig. 2, left side). Cell-mediated responses, on the other hand, involve *cytotoxic T cells* (killer T cells) and *helper T cells*. Cytotoxic T cells kill infected cells and cancer cells. Helper T cells secrete protein factors (*cytokines*), which are regulatory molecules that affect neighbouring cells. More specifically, cytokines — through helper T cells — regulate the reproduction and actions of both B and T cells. Therefore, cytokines play a pivotal role in both humoral and cell-mediated responses. Our first immune system model, presented in Section 4, incorporates most of these antibody-antigen and cell-cell interactions.

2.3 Antigen-Antibody Interaction

Antigens, such as viruses and bacteria, are mostly composed of proteins or large polysaccharides. These molecules are often part of the outer components of the coating of viruses, and the capsules and cell walls of bacteria. Generally, antibodies do not recognize an antigen as a whole molecule. Rather, antibodies identify a localized region on the surface of an antigen, called an *antigenic determinant* or *epitope*. A single antigen may have several effective epitopes, thereby potentially stimulating several different B cells to make distinct antibodies against these.

Usually, an antibody is not able to destroy an antigen directly. Rather, antibodies bind to antigens forming an antigen-antibody complex. This antigen-antibody complex provides the basis for several effector mechanisms. *Neutralization* is the most common and simplest form of inactivation, because the antibody blocks viral binding sites. The antibody will neutralize a virus by attaching to exactly those sites that the virus requires to bind to its host cell. Eventually, phagocytic cells destroy the antigen-antibody complex. This effector mechanism is part of our first simulation (Section 4).[2]

One of the most important effector mechanisms of the humoral responses is the activation of the *complement* system by antigen-antibody complexes. The complement system is a group of proteins that acts cooperatively with elements of the nonspecific and specific defense systems. Antibodies often combine with complement proteins, activating the complement proteins to produce lesions in the antigenic membrane, thereby causing lysis and hence death of the bacterial cell. *Opsonization* is a variation on this scheme, whereby complement proteins or antibodies attach to foreign cells and stimulate phagocytes to ingest those cells.

---

[2] Another effector mechanism is the *agglutination* or clumping of antigenic bacteria by antibodies. The clumps are easier for phagocytic cells to engulf than single bacteria. A similar mechanism is *precipitation* of soluble antigens through the cross-linking of numerous antigens to form immobile precipitates that are captured by phagocytes. This aspect is not considered in the IS models presented here.

Cooperation between antibodies and complement proteins with phagocytes, opsonization, and activation of the complement system is simulated in our first IS model (Section 4). We also introduce a more detailed model of the complement system in Section 11.

Another important cooperative process as part of the immune system's defense mechanisms involves macrophages. *Macrophages* do not specifically target an antigen, but are directly involved in the humoral process, which produces the antibodies that will act upon a specific antigen. A macrophage that has engulfed an antigen is termed an *antigen-presenting cell* (APC). As an APC it presents parts of the antigen to a helper T cell. This activates the helper T cell, which, in turn, causes B cells to be alerted by cytokines and consequently divide and differentiate. As a result, clones of memory B cells, *plasma B cells*, and secreted *antibodies* will be produced (Fig. 2, bottom left). These aspects are also part of our first IS model (Section 4).

3 Computational Models of the Immune System

The immune system (IS) has been studied from a modeling perspective for a long time. Early, more general approaches looked at the immune system in the context of adaptive and learning systems [16,3], with some connections to early artificial intelligence approaches [42]. Purely mathematical models, mainly based on differential equations, try to capture the overall behaviour patterns and changes of concentrations during immune system responses [43,7,19,4,48]. An algebraic model of B and T cell interactions provides a formal basis to describe binding and mutual recognition. Such algebraic models can serve as a powerful mathematical basis for further computational models, similar to formalisms for artificial neural networks [47].

Spatial aspects of immune system simulations were introduced through agent-based computational approaches in the form of cellular automata [13]. The influence of different affinities among interacting functional units, which leads to self-organizing properties, was recognized in the context of clonal selection and studied through computational models [14,2]. These models have been expanded into larger and more general simulation environments to capture various aspects of the human immune system [28,41]. There is also a large number of modeling approaches to immune system-related processes, such as for HIV/AIDS [17]. An excellent overview of these modeling strategies can be found in [18].

Most current methods consider immune response processes as emergent phenomena in complex adaptive systems [48]. Here, agent-based models play an increasingly prominent role [27,50]. This is particularly obvious in the broader application domain of bio-molecular and chemical interaction models [39]. We see the most promising potential in agent models that incorporate *swarm intelligence techniques* [6,10]. Swarm-based models result in more accurate and realistic models, in particular when spatial aspects play a key role in defining patterns of interaction, as is the case for the human immune system. Agent-based swarm models also enhance our understanding of emergent properties and help to shed some light on the inner workings of complexity as displayed by the immune system.

Biological systems inherently operate in a 3-dimensional world. Therefore, we focus our efforts on building swarm-based, 3-D simulations of biological systems. These systems exhibit a high degree of self-organization, triggered by relatively simple interactions among a large number of agents of different types. The immune system is just one example that allows for this middle-out modeling approach.[3] Other agent-based models include the study of chemotaxis within a colony of evolving bacteria [20,38], the simulation of transcription, translation, and gene regulatory processes within the lactose operon [8,22], as well as studies of affinity and cooperation among gene regulatory agents for the λ switch in $E.\ coli$ [21].

4 A Biomolecular Swarm Model: First Attempt

Our computer implementation[4] of the immune system and its visualization incorporates a swarm-based approach with a 3D visualization (Fig. 3a), where we use modeling techniques similar to our other agent-based simulations of bacterial chemotaxis, the lambda switch, and the lactose operon [22,20,38,9]. Each individual element in the IS simulation is represented as an independent agent governed by (usually simple) rules of interaction. While executing specific actions, when colliding with or getting close to other agents, the dynamic elements in the system move randomly in continuous, 3-dimensional space. This

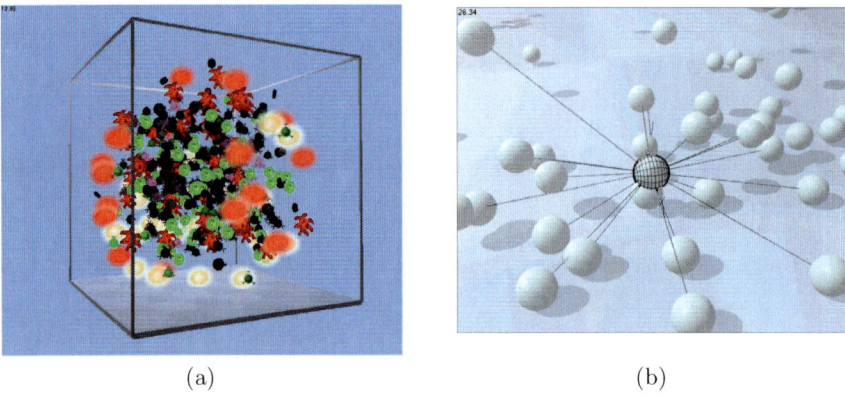

Fig. 3. Interaction space for immune system agents: (a) All interactions between immune system agents are simulated in a confined 3-dimensional space. (b) Actions for each agent are triggered either by direct collision among agents or by the agent concentrations within an agent's spherical neighbourhood space. Lines illustrate which cells are considered neighbours with respect to the highlighted cell.

---

[3] The term middle-out modeling was first coined by Sydney Brenner and recently reiterated by Denis Noble [35], where middle-out modeling provides a promising compromise and, in fact, effective connection between high-level top-down models and lower-level bottom-up modeling.

[4] We used the BREVE physics-based, multi-agent simulation engine [46].

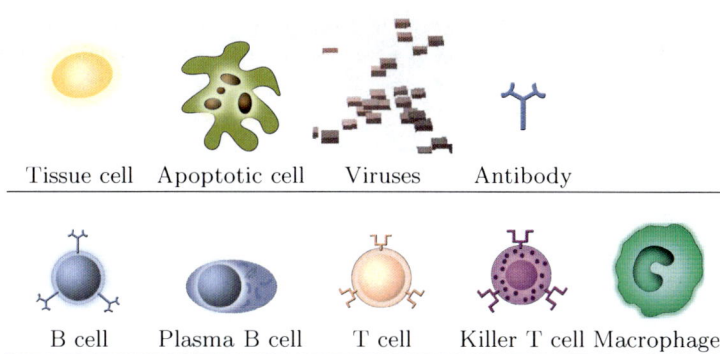

Fig. 4. The immune system agents as simulated in 3D space. The shapes have been designed to resemble typical renderings from immunology textbooks. This makes it easier to track and identify these agents in visualizations.

is different from other IS simulation counterparts, such as the discrete, 2D cellular automaton-based versions of IMMSIM [28,12]. As illustrated in Figure 4, we represent immune system agents as spherical elements of different shapes, sizes and colours. Each agent keeps track of other agents in the vicinity of its neighbourhood space, which is defined as a sphere with a specific radius. Each agent's next-action step is triggered depending on the types and numbers of agents within this local interaction space (Fig. 3b).

Confining all IS agents within a volume does, of course, not take into account that the actual immune system is spread out through a complicated

Macrophage	B Cell

`if` collision with virus:
  `if` virus is opsonized:
    Kill virus.
  `else`:
    Kill virus with prob. p.
    Create new macrophage.
 	
`if` collision with tissue cell:
  `if` cell is infected:
    `if` sufficient macrophages:
      Create new T cell.
      Create new macrophage.
	state = passive.
	`if` collision with virus:
	  state = active.
	
	`if` collision wt virus &
	active:
	  Increment vir-collision #.
	  `if` vir-collision # > TH:
	    `if` enough helper T cells:
	      Secrete antibodies.
	      Create new B cell.

Fig. 5. Simplified rules governing the behaviours of macrophages and B cells as examples of immune system agents.

network within the human body, including tonsils, spleen, lymph nodes, and bone marrow (Fig. 1). Neither do we currently—for the sake of keeping our model computationally manageable—incorporate the exchange of particles between the lymphatic vessels, blood capillaries, intestinal fluids, and tissue cells. However, we will discuss an expanded version of this model in Section 9.

Each agent follows a set of rules that define its actions within the system. As an example, we show the (much simplified) behaviours of macrophages and B cells in Figure 5. The simulation system provides each agent with basic services, such as the ability to move, rotate, and determine the presence and position of other agents. A scheduler implements time slicing by invoking each agent's Iterate method, which executes a specific, context-dependent action. These actions are based on the agent's current state and the state of other agents in its vicinity. Consequently, our simulated agents work in a decentralized fashion with no central control unit to govern the interactions among the agents.

5 Immune Response after Exposure to a Viral Antigen

We will now describe the evolution of our simulated immune response after the system is exposed to a viral antigen. Figure 6 illustrates key stages during the simulation. The simulation starts with 80 tissue cells, 14 killer T cells, 6 macrophages, 10 helper T cells, and a naive B cell. In order to trigger the immune system responses, five viruses are introduced into the simulation space (Fig. 6b). The viruses start infecting tissue cells, which turn red and signal their state of infection by going from light to dark red (Fig. 6c). The viruses replicate inside the infected cells, which eventually lyse and release new copies of the viruses, which, in turn, infect more and more of the tissue cells (Fig. 6d). The increasing concentration of viral antigens and infected tissue cells triggers the reproduction of macrophages, which consequently stimulate helper T cells to divide faster (Fig. 6e; also compare Fig. 2). The higher concentration of helper T cells stimulates more B cells and cytotoxic T cells to become active (Fig. 6f). Whenever active B cells collide with a viral antigen, they produce plasma and memory B cells and release antibodies (Fig. 6g). Viruses that collide with antibodies are opsonized by forming antigen-antibody complexes (white; Fig. 6h), which labels viruses for elimination by macrophages and prevents them from infecting tissue cells. Eventually, all viruses and infected cells have been eliminated (Fig. 7a), with a large number of helper and cytotoxic T cells, macrophages, and antibodies remaining. As all IS agents are assigned a specific lifetime, the immune system will eventually restore to its initial state, but now with a reservoir of antibodies, which are prepared to fight a second exposure to the now 'memorized' viral antigen (Fig. 7b).

The described interactions among the immune system agents are summarized in Figure 10a, which shows the number of viruses and antibodies as they evolve during the simulated humoral and cell-mediated immune response. This graph is the standard way of characterizing specificity and memory in adaptive immunity [15,45,37,1]. After the first antigen exposure the viruses are starting to

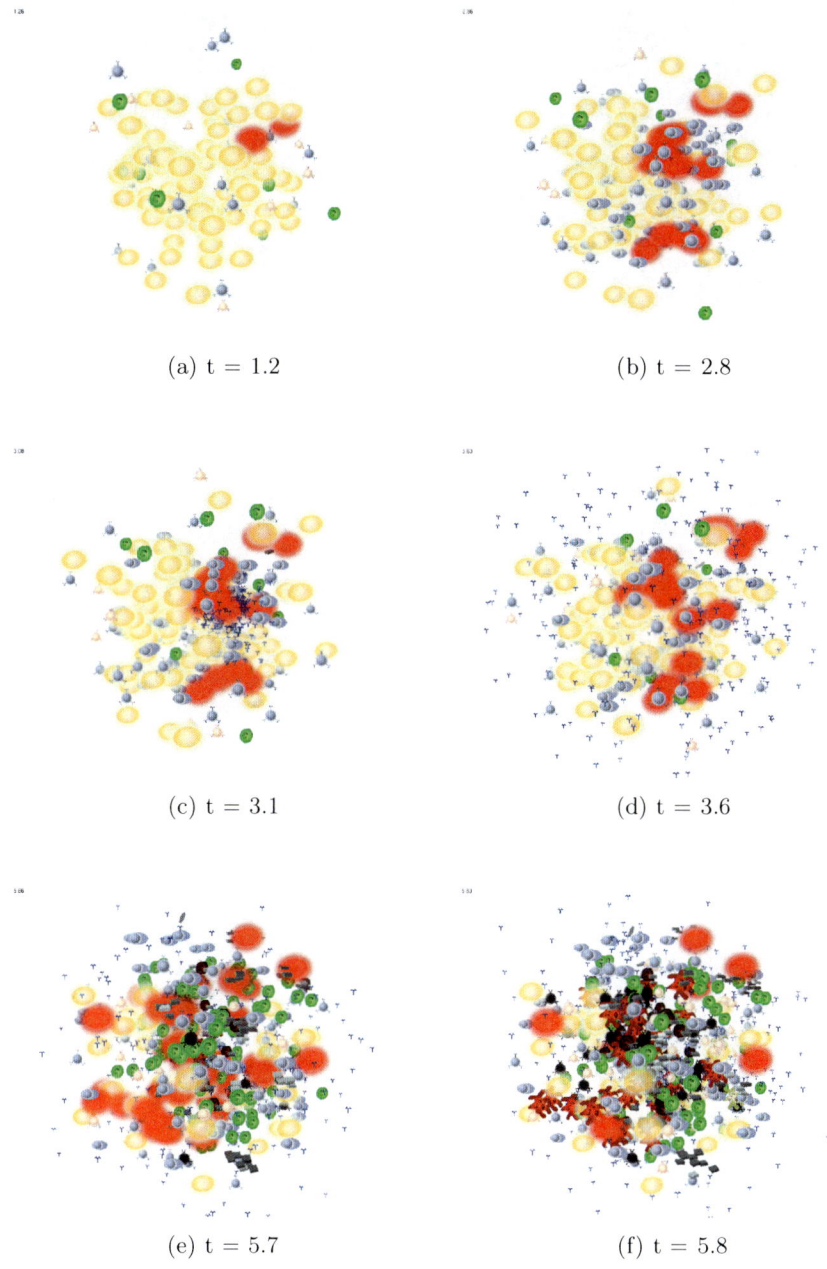

Fig. 6. Simulated immune system response after first exposure to a viral antigen.

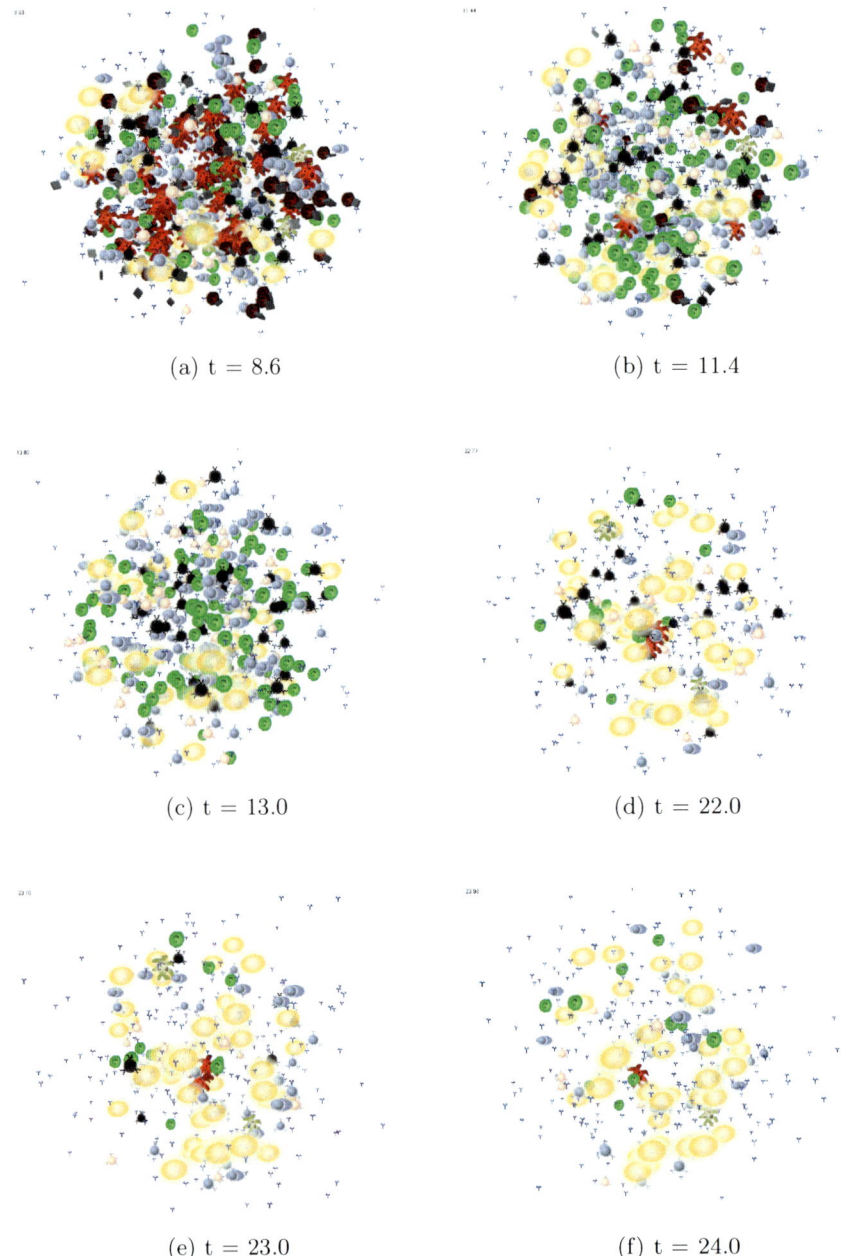

(a) t = 8.6

(b) t = 11.4

(c) t = 13.0

(d) t = 22.0

(e) t = 23.0

(f) t = 24.0

Fig. 7. Simulated immune system response after first exposure to a viral antigen (continued from Fig. 6).

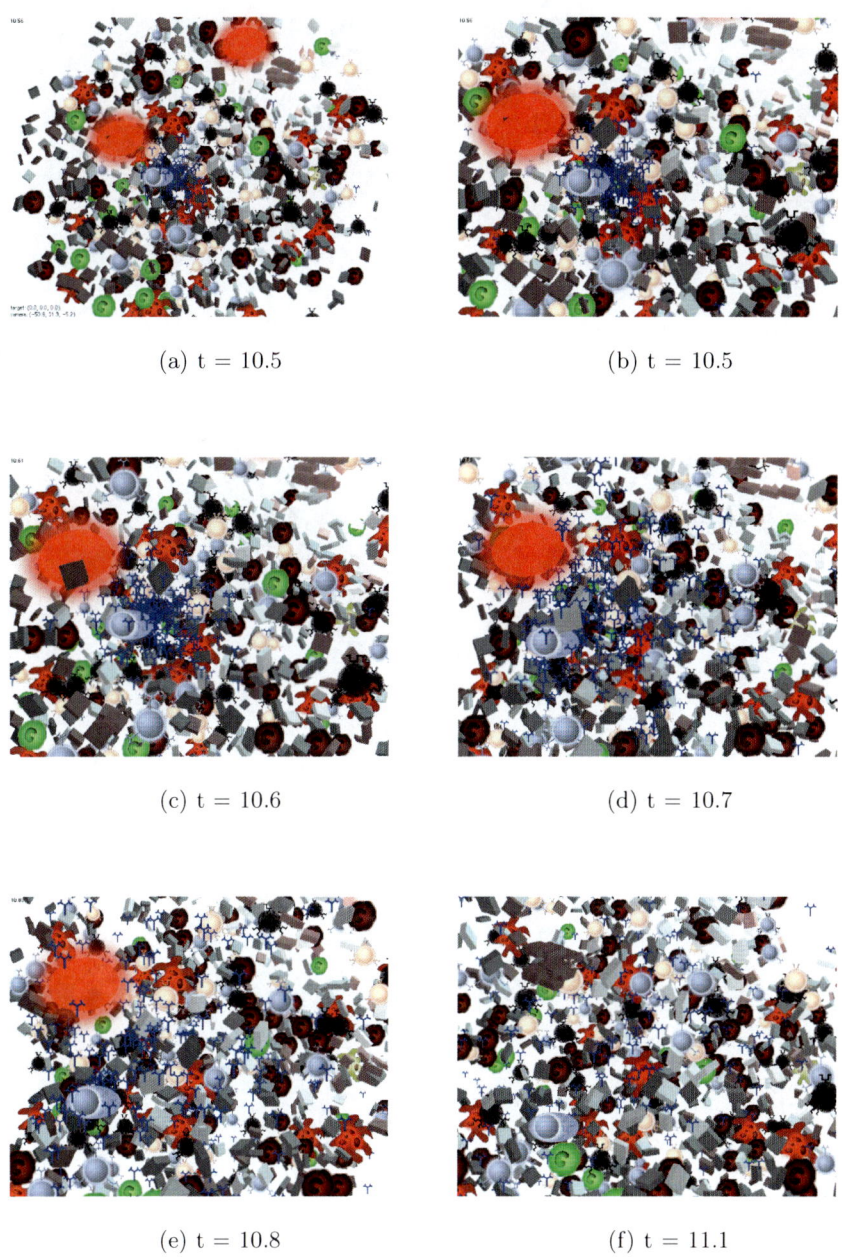

(a) t = 10.5   (b) t = 10.5

(c) t = 10.6   (d) t = 10.7

(e) t = 10.8   (f) t = 11.1

Fig. 8. A more detailed look at antibodies: Release of antibodies after collision of an activated Plasma B cell with a viral antigen. These are additional snapshots from a different viewing angle of the simulation stages shown in Figure 7(a) and (b).

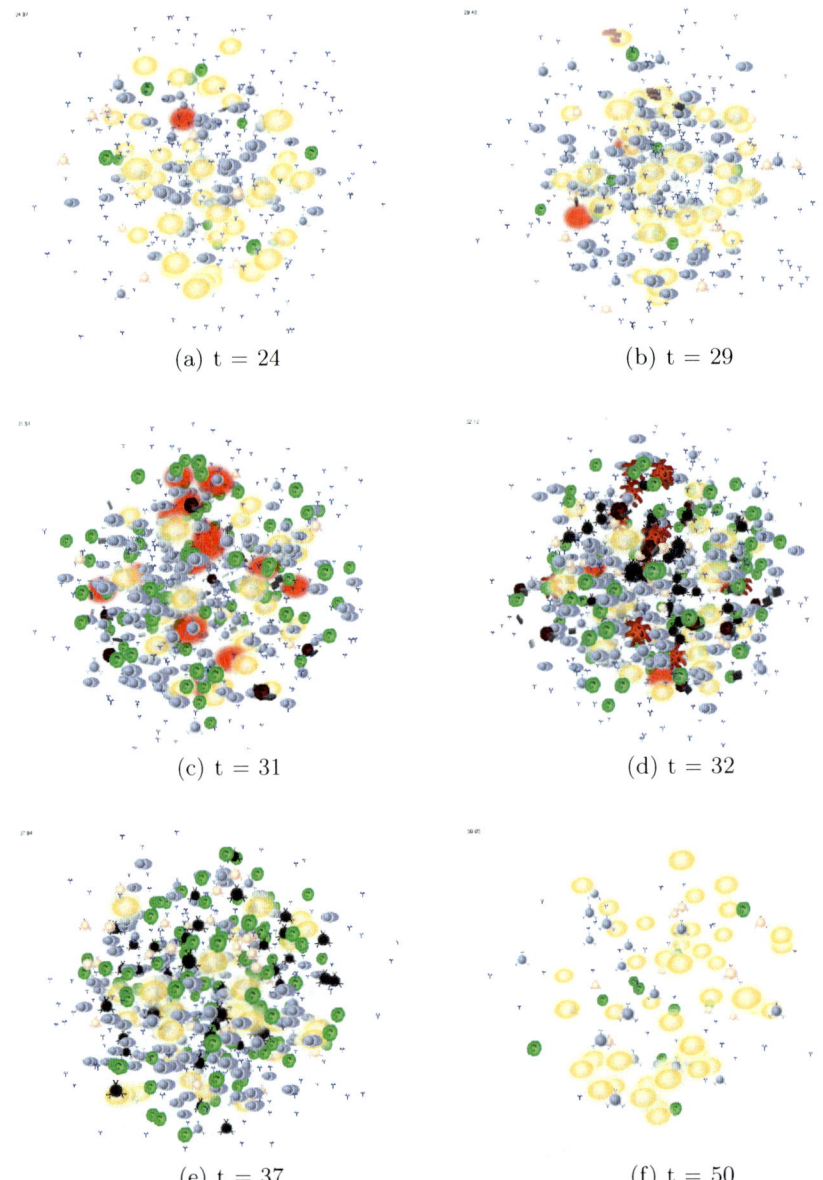

Fig. 9. Faster and more intense response after second exposure to viral antigens (continued from Fig. 7f). (a) Five viruses are inserted into the system, continuing from Step 136 after the first exposure (Fig. 7). (b) The production of antibodies now starts earlier. (c) More antibodies are released compared to the first exposure and more macrophages proliferate. (d,e) A larger number of cytotoxic T cells is produced. (f) The system falls back into a resting state, now with a 10- to 12-fold higher level of antibodies (compare Fig. 10) and newly formed memory B cells.

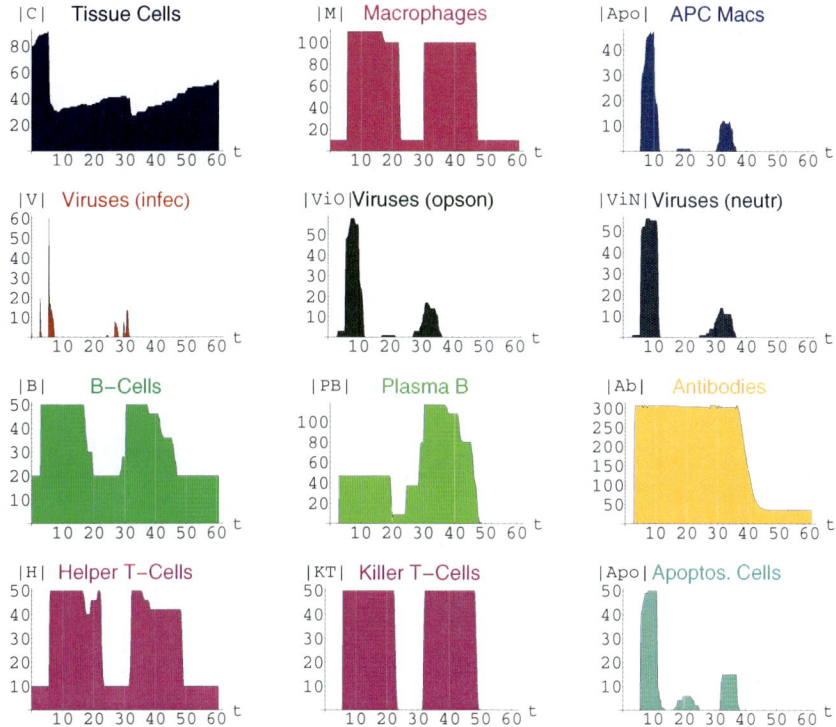

Fig. 10. Immunological Memory: The graph shows the simulated humoral immunity response reflected in the number of viruses and antibodies after a first and second exposure to a viral antigen. (a) During the viral antigen exposure the virus is starting to get eliminated around iteration $time = 70$, and has vanished from the system at $time = 90$. The number of antibodies decreases between time step 70 and 125 due to the forming of antigen-antibody complexes, which are then eliminated by macrophages. A small amount of antibodies (10) remains in the system. (b) After a second exposure to the viral antigen at $t = 145$, the antibody production is increased in less than 50 time steps. Consequently, the virus is eliminated more quickly. About 13 times more antibodies (130) remain in the system after this second exposure (reproduced from [23]).

get eliminated around iteration $time = 50$, and have vanished from the system at $time = 100$. The number of antibodies decreases between time step 50 and 100 due to the forming of antigen-antibody complexes, which are eliminated by macrophages. Infected tissue cells are lysed by cytotoxic T cells, which delete all cell-internal viruses. After all viruses have been fought off, a small amount of antibodies remains in the system, which will help to trigger a more intense and faster immune response after a second exposure to the same antigen, which is described in the following section.

6 Immune System Response after Second Exposure to Antigen

The selective proliferation of lymphocytes to form clones of effector cells upon first exposure to an antigen constitutes the *primary immune response*. There is a lag period between initial exposure to an antigen and maximum production of effector cells. During this time, the lymphocytes selected by the antigen are differentiating into effector T cells and antibody-producing plasma cells. If the body is exposed to the same antigen at some later time, the response is faster and more prolonged than during the primary response. This phenomenon is called the *secondary immune response*, which we will demonstrate through our simulated immune system model (Fig. 10b).

The immune system's ability to recognize a previously encountered antigen is called *immunological memory*. This ability is contingent upon long-lived memory cells. These cells are produced along with the relatively short-lived effector cells of the primary immune response. During the primary response, these memory cells are not active. They do, however, survive for long periods of time and proliferate rapidly when exposed to the same antigen again. The secondary immune response gives rise to a new clone of memory cells as well as to new effector cells.

Figure 9 shows a continuation of the immune response simulation of Figure 7. We introduce five copies of the same virus the system encountered previously. We keep track of all viruses inserted into the system and can thus reinsert any previous virus, for which antibodies have been formed. Each virus, which is introduced into the system, receives a random signature $s \in [0, 10]$. Once memory B cells collide with a virus, they produce antibodies with the same signature, so that those antibodies will only respond to this specific virus. After a second exposure to the same viral antigen, the highest concentration of antibodies is increasing (Fig. 10). Consequently, the virus is eliminated much faster, as more antigen-antibody complexes are formed, which get eliminated quickly by the also increased number of macrophages. Additionally, an increased number of helper and killer T cells contributes to a more effective removal of infected cells (Fig. 9). Not even half the number of viruses can now proliferate through the system, compared to the virus count during the first exposure. After the complete elimination of all viruses, ten to fifteen times more antibodies remain in the system after the second exposure. This demonstrates that our agent-based model—through emergent behaviour resulting from agent-specific, local interaction rules—is capable of simulating key aspects of both humoral and cell-mediated immune responses.

7 Lessons Learnt from Model One

We think that a decentralized swarm approach to modelling the immune system closely approximates the way in which biologists view and think about living systems. Although our simulations have so far only been tested for a relatively small number of interacting agents, the system is currently being expanded to handle a much larger number of immune system agents and other biomolecular entities (such as cytokines), thus getting closer to more accurate simulations of massively-parallel interaction processes among cells that involve hundreds of thousands of particles.

A swarm-based approach affords a measure of modularity, as agents can be added and removed from the system. In addition, completely new agents can be introduced into the simulation. This allows for further aspects of the immune system to be modelled, such as effects of immunization through antibiotics or studies of proviruses (HIV), which are invisible to other IS agents.

8 Decentralized Information Processing

One aspect that makes the human immune system particularly interesting—but more challenging from a modeling perspective—is its vastly decentralized arrangement. Tissue and organs of the lymphatic system are widely spread throughout the body, which provides good coverage against any infectious agents that might enter the body at almost any location. Even the two key players responsible for specific immunity originate from different locations within the body: T cells come from the thymus, whereas B cells are made in the bone marrow. The lymphocytes then travel through the blood stream to secondary lymphoid organs: the lymph nodes, spleen, and tonsils. Within these organs, B and T cells are rather tightly packed, but can still move around freely, which makes them easier to model as agents interacting in a 3-D simulation space.

Lymph nodes can then be considered the primary locations of interactions among T cells, activated by antigen presenting cells. T cells, in turn, activate B cells, which evolve into memory B cells and antibody-producing plasma B cells. Both types of activated lymphocytes will subsequently enter the lymphatic system, from where they eventually return to the blood stream. This enables the immune system to spread its activated agents widely throughout the body. Finally, the lymphocytes return to other lymph nodes, where they can recruit further agents or trigger subsequent responses. Hence, B and T cells as well as other immune system agents (antibodies, cytokines, dendritic cells, antigen presenting cells, etc.) are in a constant flow between different locations within the human body [36].

9 Simulating Decentralized Immune Responses

Our overall goal is to build a whole body simulation of the immune system (Fig. 11). This, of course, does not only require a large amount of computing resources, but also requires a modular and hierarchical design of the simulation framework. Modelers – i.e., immunologists as well as researchers and students in health sciences – should be able to look at the simulated immune system at different levels of detail. A whole body simulation will not be as fine grained as when looking at the interactions within a lymph node or at the intersection between the lymphatic and vascular system. In our current implementation we have incorporated three distinct, but interconnected sites within the human body that are related to the immune system:

- **Lymph Nodes:** Within a lymph node section we incorporate adaptive immune system processes during clonal selection, in response to viral antigens entering the lymph node. Different types of B cell strands can be defined. In case of a high degree of matching with an antigen, rapid proliferation is triggered.
- **Tissue:** Within a small section of tissue we model the immune system processes during primary and secondary response reactions among viruses (with their associated antigen components), tissue cells, dendritic cells, helper T and killer T cells, memory and plasma B cells (with their associated antibodies), and macrophages.

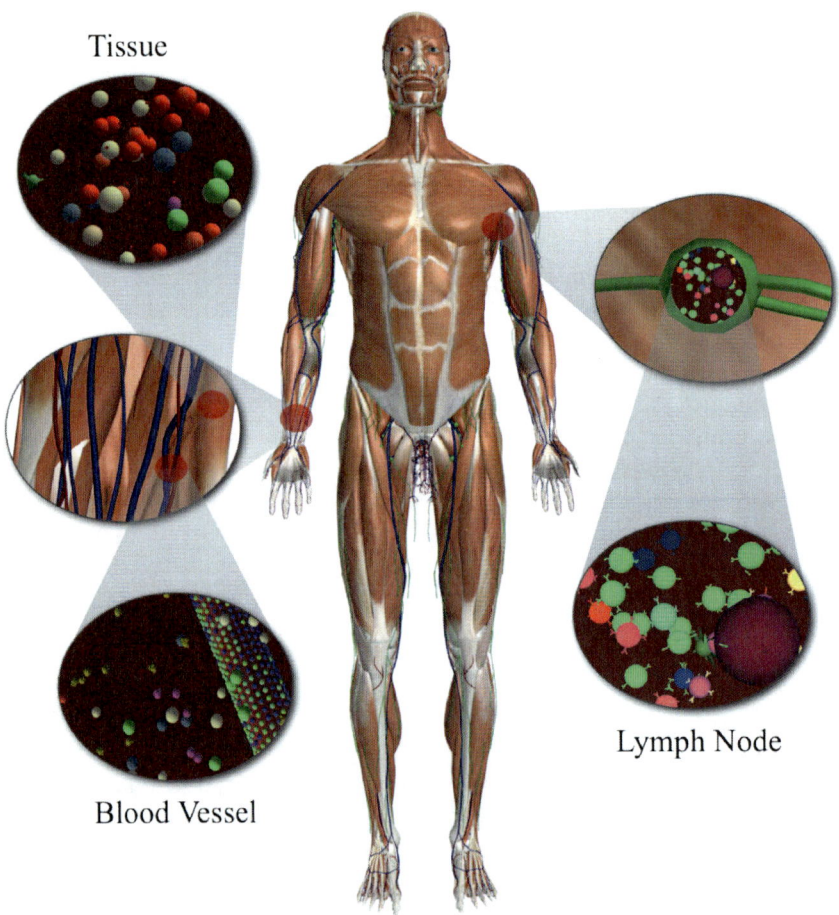

Fig. 11. The decentralized defenses of immunity. Three compartmental modules, that exhibit distinct but interconnected functionalities within the human immune system, are implemented in our immune system simulations: tissue, blood vessels, and lymph nodes.

- **Blood Vessel-Tissue Interfaces:** At the interface between blood vessels and tissue, we simulate red blood cells moving within a section of a blood vessel, lined with endothelial cells, which can produce selectin and intercellular adhesion molecules (ICAMs). This causes neutrophils to start rolling along the vessel wall and exit the blood stream into the tissue area. Any bacterium within the tissue is subsequently attacked by a neutrophil. During ingestion of a bacterium by a macrophage, tumor necrosis factor (TNF) is secreted and the bacterium releases lipopolysaccharides (LPS) from its surface. In turn, TNF triggers selectin production in endothelial cells, whereas LPS induces endothelial cells to produce ICAM.

The following sections explain our model in more detail with respect to clonal selection as well as primary and secondary responses within a lymph node area and a tissue region (Section 9.1). The IS processes triggered during a bacterial infection within the interface area between a blood vessel and tissue is described in Section 9.2.

9.1 Simulated Viral Infection

Figure 12 gives an overview of the immune system agents and their interaction patterns in our model. Each agent is represented by a specific, 3-dimensional shape, which are also used in the (optional) visual representation of the agents during a simulation experiment. We demonstrate one experiment to show a typical simulation sequence.

Clonal Selection within a Lymph Node: In this experiment, we first focus our attention on a selected lymph node in order to observe the IS agent reactions after a virus enters the lymph node area (cf. Fig. 11). Initially, 50 B cells as well as 20 helper-T cells of 8 different types (signatures) are present. Figure 13f shows that there is a fairly even initial distribution of the different strands of B and T cells. Around time step $t = 14.6$, dendritic cells enter the lymph node and present a single type of viral antigen (Fig. 13b), which stimulates a nearby helper-T cell and causes a matching B cell (following the Celada-Seiden affinity model [13]) to replicate. Soon after ($t = 57.1$), a significantly larger population of matching B cells proliferates the lymph node area (Fig. 13c), where B cells have already started to emit antibodies. In Fig. 13f the concentration of these fast proliferating B cells is represented by the green plot. At time point $t = 225.0$, memory B cells of the matching strand have become more common. Around $t = 256.4$, the same virus is introduced into the lymph node again. Now it is mainly the memory B cells that trigger the secondary response and replication of plasma B cells, which secrete antibodies (compare the increase of the matching B cell concentration (green) towards the last third of the graph in Fig. 13f).

Primary and Secondary Response in Tissue: At the same time, while the simulation of the interactions within the lymph node are running, a concurrent, second simulation models the response processes in a selected tissue area (cf.

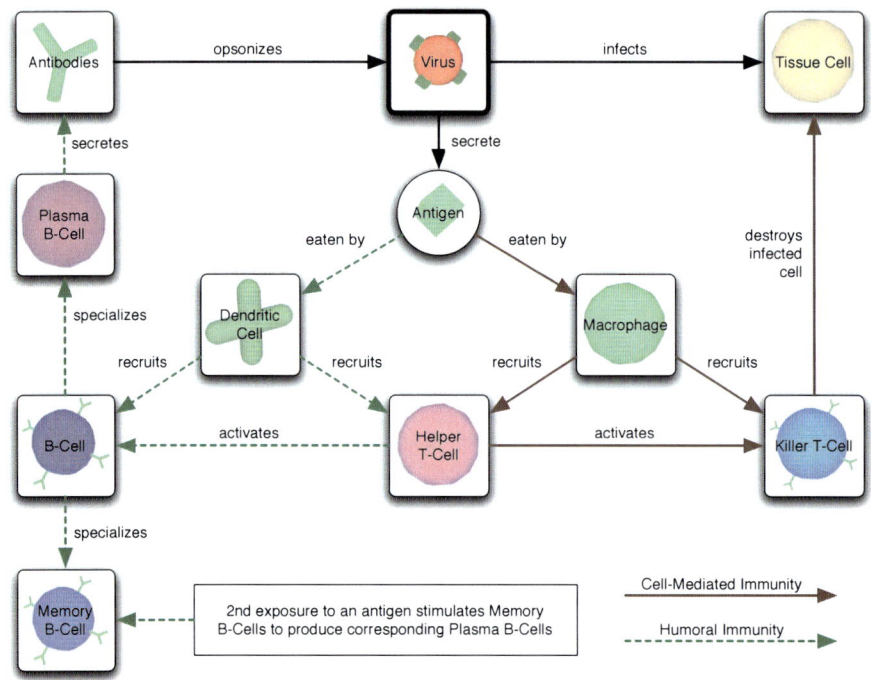

Fig. 12. Interactions of immune system agents triggered by viral infection: A virus is usually identified by its antigens, which alert both dendritic cells and macrophages to ingest the viruses. Both actions lead to recruitment of further IS cells. Dendritic cells recruit B cells, which – in particular when activated by helper-T cells – replicate as memory B cells or proliferate into plasma B cells, which in turn release antibodies to opsonize the virus. On the other hand, macrophages with an engulfed virus stimulate an increase in the proliferation of both helper and killer T cells, which are the key players in cell-mediated immunity and destroy virus-infected tissue cells to prevent any further spreading of the virus (reproduced from [25]).

Fig. 11). Circulation of IS agents is implemented by a communication channel between lymph node and tissue areas. Within the tissue simulation space (Fig. 14a), we start with 10 dendritic cells, 5 killer-T cells, 5 helper-T cells, 5 macrophages, 60 tissue cells and 5 copies of the same virus introduced into the lymph node as described above. In Fig. 14b a cell has been infected by the virus and antibodies (from the lymph node) start entering the tissue area. Figure 14c shows a close-up of the important agents: one virus is visible inside an infected cell, another virus has docked onto the surface of a tissue cell and is about to enter it. A third virus has already been opsonized by an attached antibody. Now macrophages will start to engulf opsonized viruses and more macrophages are recruited in large numbers (Fig. 14d). This triggers an analogous spike in the number of killer-T and helper-T cells (compare Fig. 15). The increase in killer-T

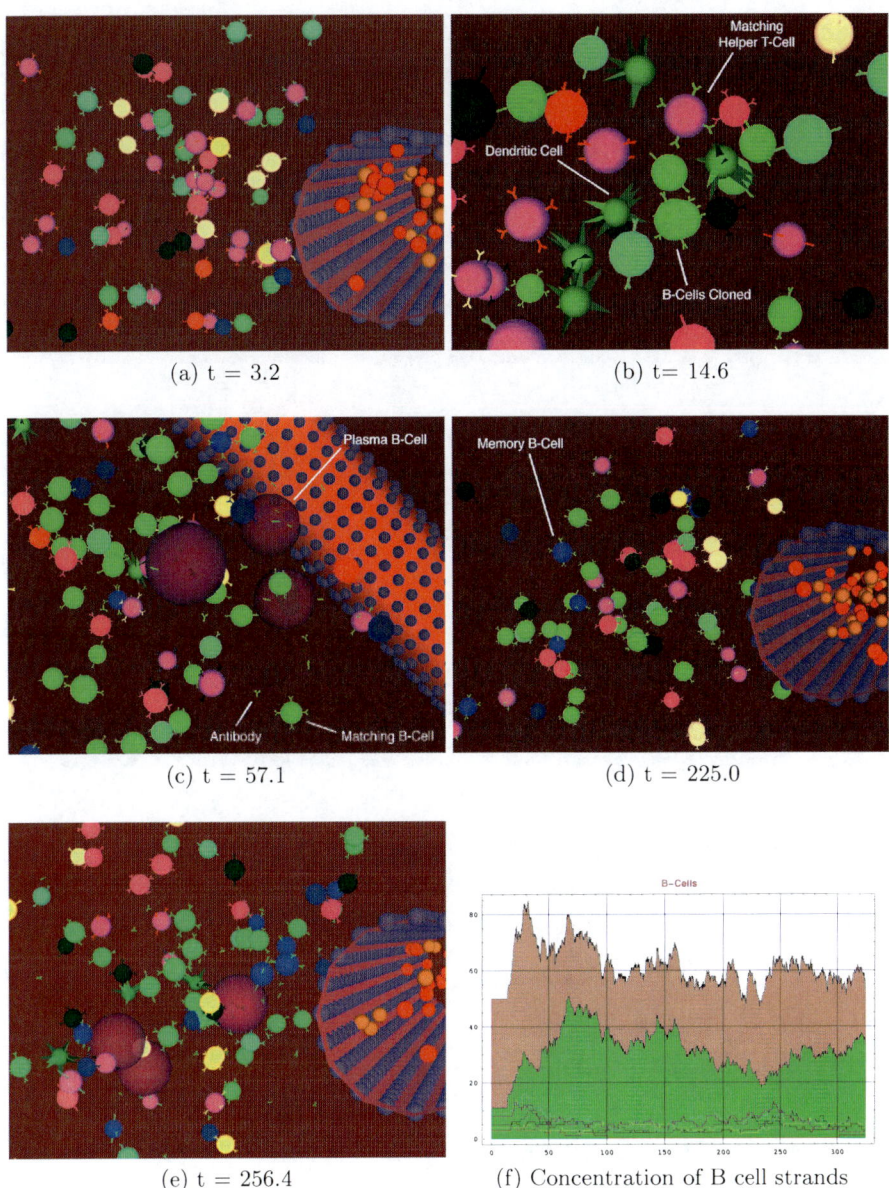

Fig. 13. Interactions in a **Lymph Node** after a viral infection: (a)-(e) Screen captures (with time point labels) of the graphical simulation interface during clonal selection and primary and secondary response to a virus. The virtual cameras are pointed at a lymph node, in which 8 different strands of B cells are present. (f) The change in concentration of all B cells (brown filled plot) and per strand. The virus that most closely matches one of the B cell strands triggers its increased proliferation (green filled plot). The concentrations of all other strands remain low (line plots at the bottom). This figure is reproduced from [25]

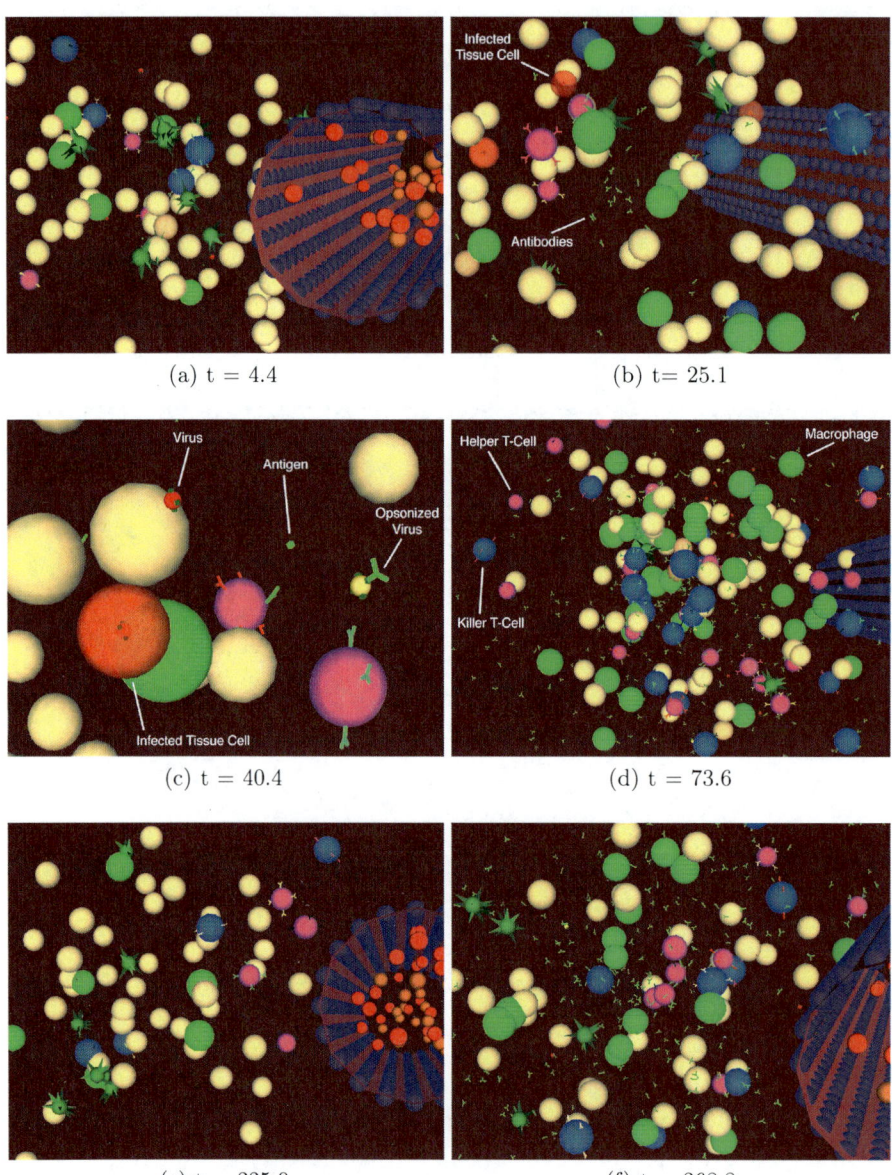

Fig. 14. Interactions in a **Tissue Area** after a viral infection: Screen captures of the graphical simulation interface during clonal selection and primary and secondary response after viral infection. The virtual cameras are pointed at a tissue region close to a blood vessel (reproduced from [25]).

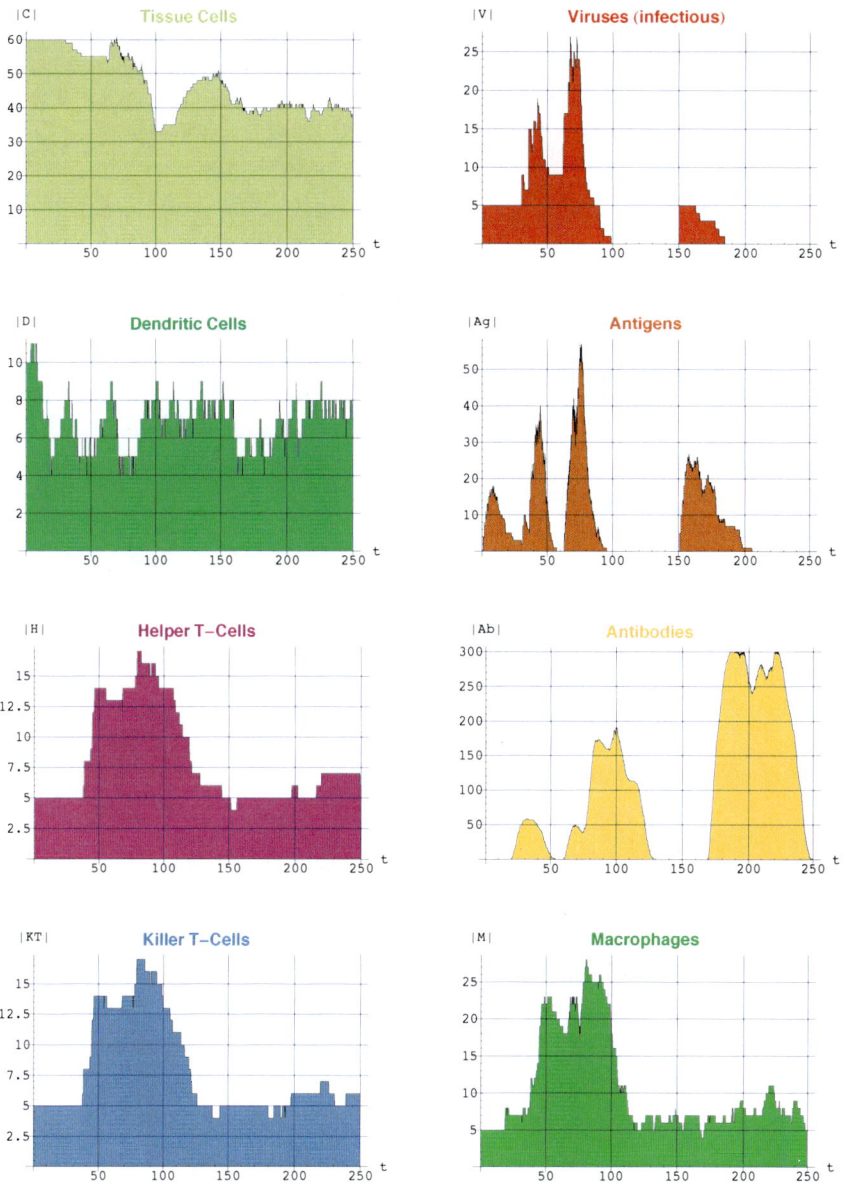

Fig. 15. Evolution of IS agent concentrations during the primary and secondary responses in a tissue area (reproduced from [25]).

cells makes it more likely for these cells to collide with an infected tissue cell and initiate its apoptosis.

After about 120 time steps, the infection has been fought off, with no more viruses or antigens remaining in the system (Fig. 15). The concentrations of

T cells and macrophages return to their initial levels. At $t = 150.0$, the same virus is reinserted into the system. Memory B cells inside the lymph node create an influx of plasma B cells almost immediately. As a result, the infection is stopped within a much shorter time interval, due to the increased amount of antibodies. Cell-mediated immunity reactions do start faster as well, but are not as intense as during the first response, since the infection is eliminated more quickly. Consequently, T cells and macrophage concentrations can remain at a lower level.

9.2 Simulated Bacterial Infection

In addition to viral infection, we now look more closely at a simulation of infections caused by bacteria and how the immune system tries to keep bacterial infections in check. Bacteria multiply within the tissue. Their waste products, produced from a large concentration of bacteria, can be damaging to the human body. Therefore it is important that the immune system kills off bacterial invaders before a critical concentration is reached.

The following experiment demonstrates immune system response processes during bacterial infection. The key players and their interactions are outlined in Fig. 16. As this involves not only bacteria and macrophages but also neutrophils that enter tissue from the vascular system, the simulation space comprises a segment of a blood vessel (Fig. 17). The tissue-vessel interface area is initialized with tissue cells, B cells, helper-T cells, macrophages, and a number of bacteria acting as infectors. The blood vessel, lined with endothelial cells, contains red blood cells and neutrophils.

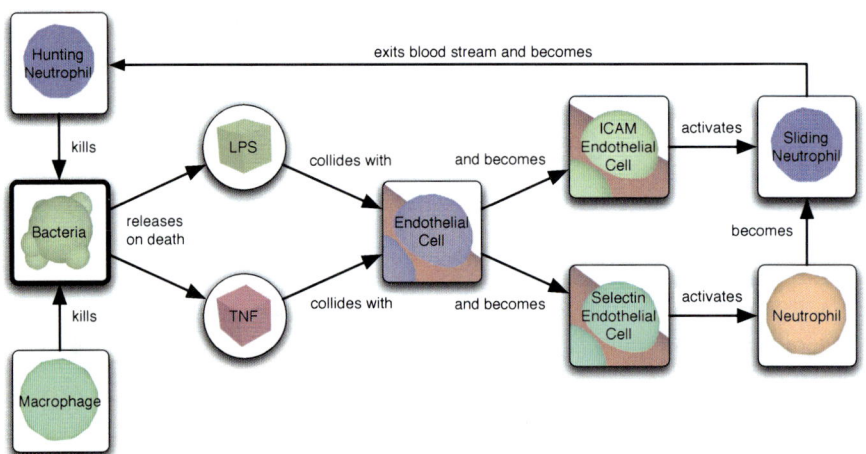

Fig. 16. Bacterial Infection: A summary of the interaction network between bacteria, macrophages, neutrophils, and endothelial cells that line the blood vessel (reproduced from [25]).

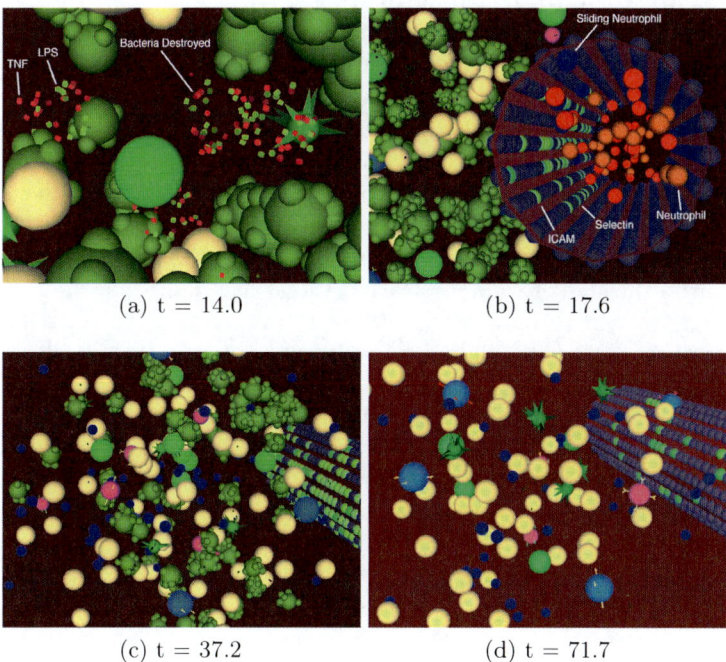

Fig. 17. Fighting Bacterial Infection: (a) macrophages attacking bacteria, (b) endothelial cells, neutrophils and red blood cells inside the blood vessel, (c) neutrophils (blue) on their hunt for bacteria, (d) all bacteria have been eliminated (reproduced from [25]).

Macrophages that engulf bacteria release TNF (tumor necrosis factor), while lipopolysaccharides (LPS), which are major structural components of Gramnegative bacterial cell walls, are released into the tissue area (Fig. 17a). Once endothelial cells get in contact with TNS or LPS, they release selectin or intercellular adhesion molecules (ICAMs), respectively (Fig. 17b). When a neutrophil collides with an endothelial cell that produces selectin, it will start to roll along the interior surface of the blood vessel. A neutrophil rolling along an ICAM-producing endothelial cell will exit the blood stream and head into the tissue area. Once in the tissue area, neutrophils—together with macrophages— act as complementary hunters of bacteria (Fig. 17c). Notice the high number of activated endothelial cells in the blood vessel wall. A bacterium colliding with a neutrophil is engulfed and consumed, while LPS and TNF are again released into the system. Finally, all bacteria have been eliminated and the number of activated endothelial cells is reduced (Fig. 17d). Neutrophils will soon disappear since the system has recovered from the bacterial infection.

10 Lessons Learnt from Model Two

Our simulation environment is currently used as a teaching tool in biology, medical, and computer science undergraduate and graduate classes. Due to its visual interface and the ability to specify many simulation control parameters through configuration files, it serves both as an educational device and as an exploration tool for researchers in the life sciences. Students seem to gain a more 'memorable' understanding of different aspects of immune system processes. Although visualizations can also be misleading, they usually help in grasping essential concepts, in particular in the case of an orchestrated system of a multitude of agents, such as the immune system. From our experience, the visualization component is important for a proper understanding of emergent processes that result from the interplay of a relatively large number of agents of different types with simple but specific local interaction rules. Gaining an understanding and 'intuition' about emergent properties as in the immune system plays a key role in building today's biologically accurate computer simulations.

Of course, our current version does not even come close to the actual numbers of interacting IS agents (e.g., billions of B cells within a small lymph node section). However, according to our experience, key effects within an agent-based interaction system can already be observed with much smaller numbers. Usually, only a 'critical mass' is needed. This is certainly an area that requires further investigation. Using evolutionary computation techniques, we also explore the effects of different control parameter settings, as well as how changes in the set of agent interaction rules influence the overall system behaviour [30,32]. Furthermore, being able to easily change agent interaction rules and agent types makes models of complex adaptive systems useful for large-scale scientific exploration [24,31].

11 A Swarm Model of the Complement System

The complement system is one of the most ancient forms of immunity. Nearly 700 million years old, it predates the emergence of antibodies by about 250 million years, long before the evolution of adaptive immunity [44]. In its current evolution, it serves an important role in innate immunity. The complement system also forms a bridge to adaptive immunity [26] through interactions with antibodies and various anaphylotoxin messenger peptides that alert other branches of immunity into action. Without a healthy and functioning complement system, an organism's survival is severely compromised [49,5].

The human complement system can be subdivided into three different pathways: Classical, Alternative and Lectin (Fig. 18). All three pathways converge on a common point at C3b, which is nearly the endpoint of the *Early Complement* period [29]. As the Lectin pathway operates similar to the Classic pathway, we only consider the Classical and Alternative pathways in our model discussed here.

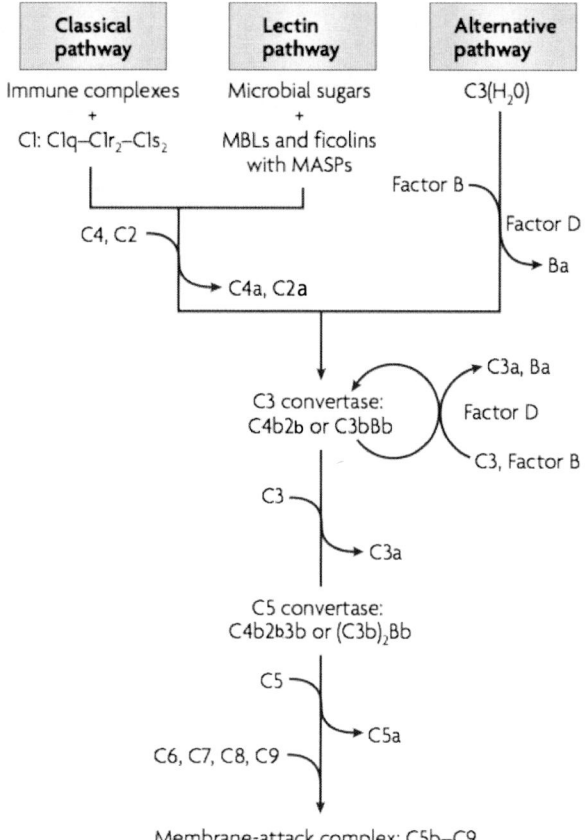

Fig. 18. The complement system is conceptually subdivided into three pathways.

The *Classical pathway* begins with antigen-antibody complexes forming on the surface of a pathogen, such as a bacterium. C1 protein binds with these antibodies and causes the activation of serine proteases in the head. These can then convert C4 to C4b (and C4a). C4b then binds to the pathogen surface and facilitates the binding and conversion of C2 to C2b (and C2a). The combination of C4b:C2b results in a C3-Convertase, which can convert C3 into C3b (and C3a). A free C3b can also associate with the C3-Convertase to give C3b:C4b:C2, which is a C3/C5 Convertase.

The *Alternative pathway* is unique in that it does not require any antibody interaction to target a pathogen. Bio-agents of the Alternative pathway accomplish this indirectly in two different ways. First, B-Protein can bind to already bound C3b. Once activated by D-Protein, C3b can generate a C3-Convertase. Similar to the Classic pathway, the incorporation of another C3b, to give C3b:C3b:B(activ), results in a C3/C5 Convertase. The second way the Alternative pathway can interact with a pathogen is by a fluidphase C3 Convertase. As it turns out,

(a) (b)

Fig. 19. (a) The basic pathogen agent environment. (b) An example of all the agents displayed in the system while a simulation is running.

approximately 0.5 percent of serum C3 has undergone hydrolysis into C3(H_2O). This allows B-Protein to associate with it, which is then activated by D-Protein; the result is a fluid-phase C3 Convertase. The feedback with the generation of C3b of the Alternative pathway contributes positively to an amplification effect on the Classic pathway. Indeed, the Alternative pathway is considered to be a type of amplification pathway [26].

11.1 A Swarm Model of the Early Complement

The experimental environment[5] consists of a square box and a single pathogen in the centre. The remainder of the agents in the system interact in the region between the pathogen and the barrier walls (Fig. 19). The size of the box and the pathogen radius are easily accessed variables in the model. This Swarm uses a modified interaction model of the complement system as the bases of the agent interaction rules (Fig. 20). It should be noted that the model does not create or track C4a, C2a, C3a or C5a in the simulation. However, the amount of C4b, C2b, C3b and C5b can give an indication of the numbers of a-peptides generated in the system. Since regulatory proteins are not yet included in the model, decay rates on some of the proteins and complexes are used instead. The model generates a reaction up to the activation of C5 to C5b (and C5a). As the formation of the MAC complex is considered to be quite fast, a collision of C5b with the pathogen surface is considered to indicate a MAC formation. The model can also be easily adjusted to run one pathway at a time, or both pathways at once. At every interval iteration, activation times for all complement proteins are recorded. Formation of the various convertases in each pathway and the associated C3b, C5b generation, is also recorded and can be accessed individually at anytime during the experiment.

---

[5] We used the BREVE physics-based, multi-agent simulation engine [46] for the implementation of this model.

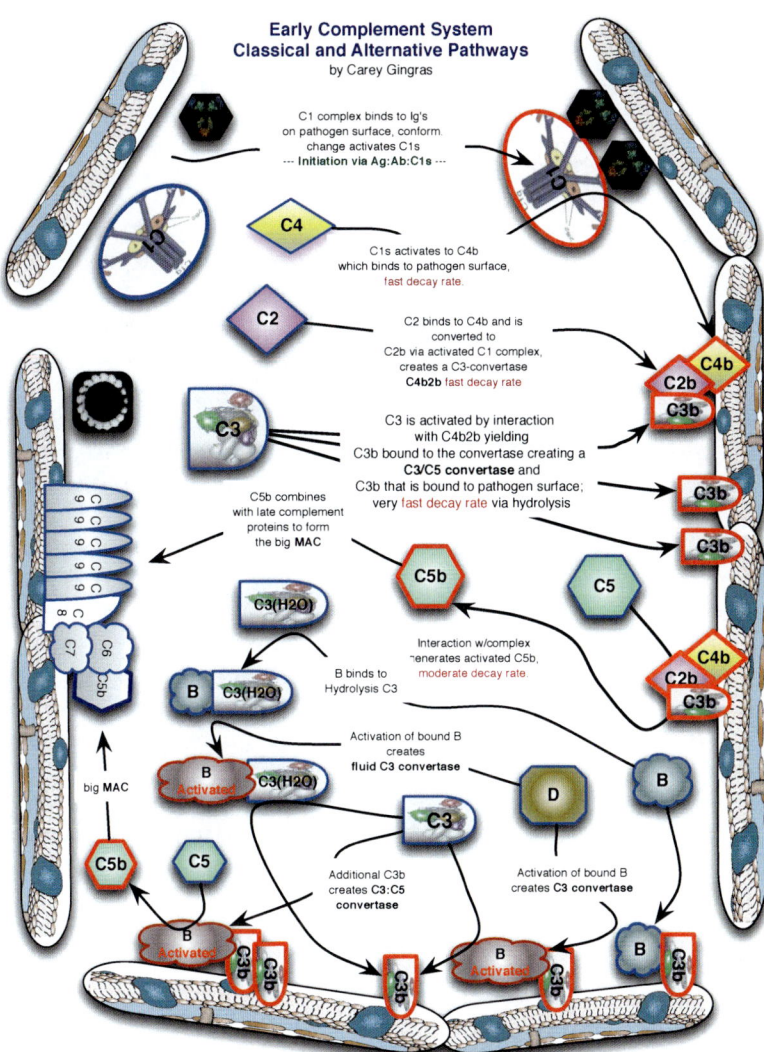

Fig. 20. The interaction model for the three different Complement Pathways as used in our model.

11.2 Results from the Complement Model

Experimental runs were performed for 8000+ iterations for both the Classical and Alternative Pathways, and each individually. All agents are regenerative, which means we assume a constant concentration of complement proteins around a single pathogen. The starting concentrations for each type of bio-agent are shown in Fig. 21. What is first noticeable about the visual results

Complement Protein	Agent Population
Ab	20
C1	34
C4	248
C2	30
C3	560
C5	36
B	40
D	50
Pathogen	1

Fig. 21. Initial Agent Populations (1008 bio-agents in total)

(Fig. 22), is the minimal protein deposits on the pathogen surface of the Alternative pathway alone (Fig. 22(a),bottom). Compared to the Classic and combined Classic-Alternative pathways, the Alternative pathway, by itself, invokes a very protracted response; it would take much longer to fully tag the pathogen. Within 8000 world time units, both the Classical and combined runs result in a relatively thorough coating of the pathogen surface.

Examining Fig. 23, we see slightly different agent profiles for the Classic and combined runs. These graphs show all the agent populations and their rates of generation over time. While the rate of C3b deposited on the pathogen surface in both cases is similar, the total number of C3b created is about 10 percent higher than in the combined pathway. Another noticeable difference between the two runs is in the initial pattern in C3b generation. When both pathways are run at the same time, most of the initial C3b production comes from the Alternative pathway components, whereas C3b from the Classical pathway does not start to increase production until 2000 world iterations. This is in contrast to the classical pathway, where production begins to ramp up around 1000 iterations. Overall, both pathways for the entire production run generate about the same number of C3b: 1862 in the Classic pathway and 1963 for both pathways. However, the C3b production from the Classic components, when both are run at the same time, appears to be delayed in the first quarter of the experiment — or "shifted to the right" when the Alternative pathway agents are added.

C3b pathogen binding by the combined pathways is 5 percent greater than the Classic pathway alone. However, the amount of C4b pathogen binding is 13 percent lower than the Classic run by itself. In both cases, the total number of C4b bound levels off to about 220 agents for a pathogen of this size. Antibody binding also levels off at about 55, as does C1 binding, which parallels the amount of antibodies bound.

Looking at convertase production, in the case of the Classic pathway, 6 C3 convertases and 173 C3:C5 convertases were formed. In the run with combined pathways, 212 C3 convertases and 187 C3:C5 convertases were created. Generally, we achieve a 10 to 12 percent increase when both pathways are run together. In addition, both experimental runs produced approximately the same number

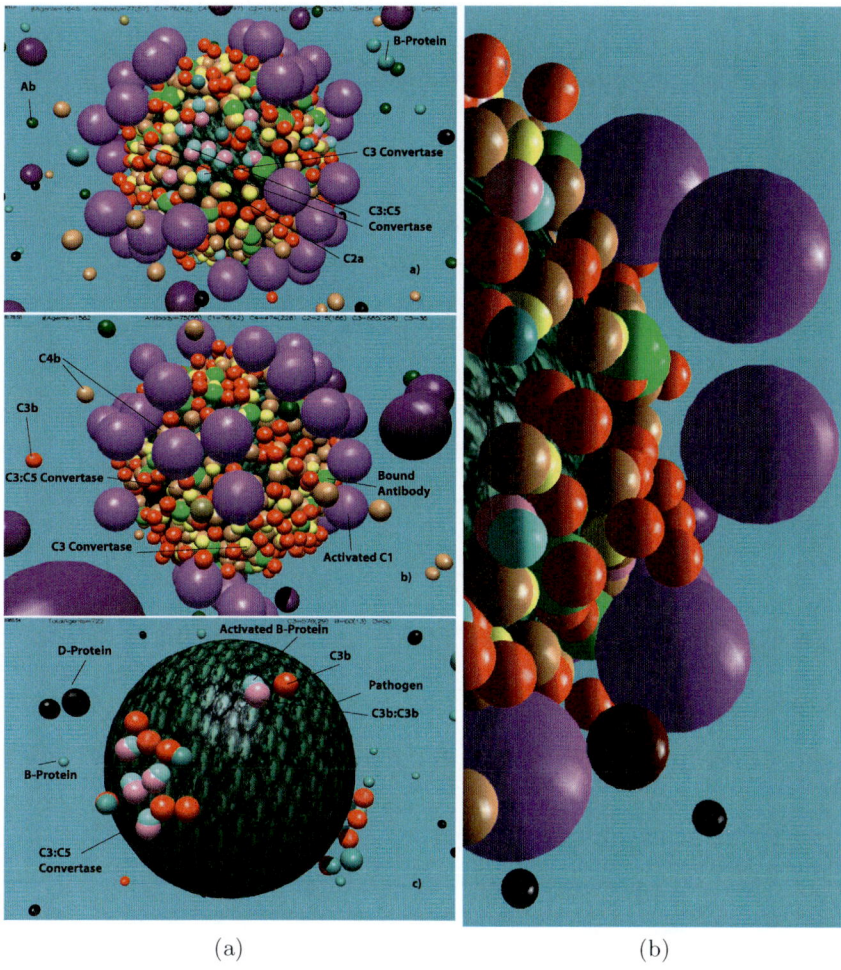

Fig. 22. Close-ups of the Complement System simulation: (a) The Classic and Alternative Pathways, the Classic Pathway, and the Alternative Pathway (from top to bottom; all displayed after approximately 8000 simulation steps), (b) a section of the pathogen surface at 3500 iterations.

of activated C5b with potential impact for the formation of the membrane attack complex (MAC). In the Classic pathway alone, 186 Cb5 were created, with 12 potential MAC attacks. For the combined pathways, 101 C5bs were created and 11 potential MAC attacks.

11.3 Evaluation of the Complement Model

While Classic pathway proteins were calibrated to previously measured serum levels [34], the Alternative protein concentrations still need to be adjusted. When this is accomplished, a more valid comparison of the two pathways can be made.

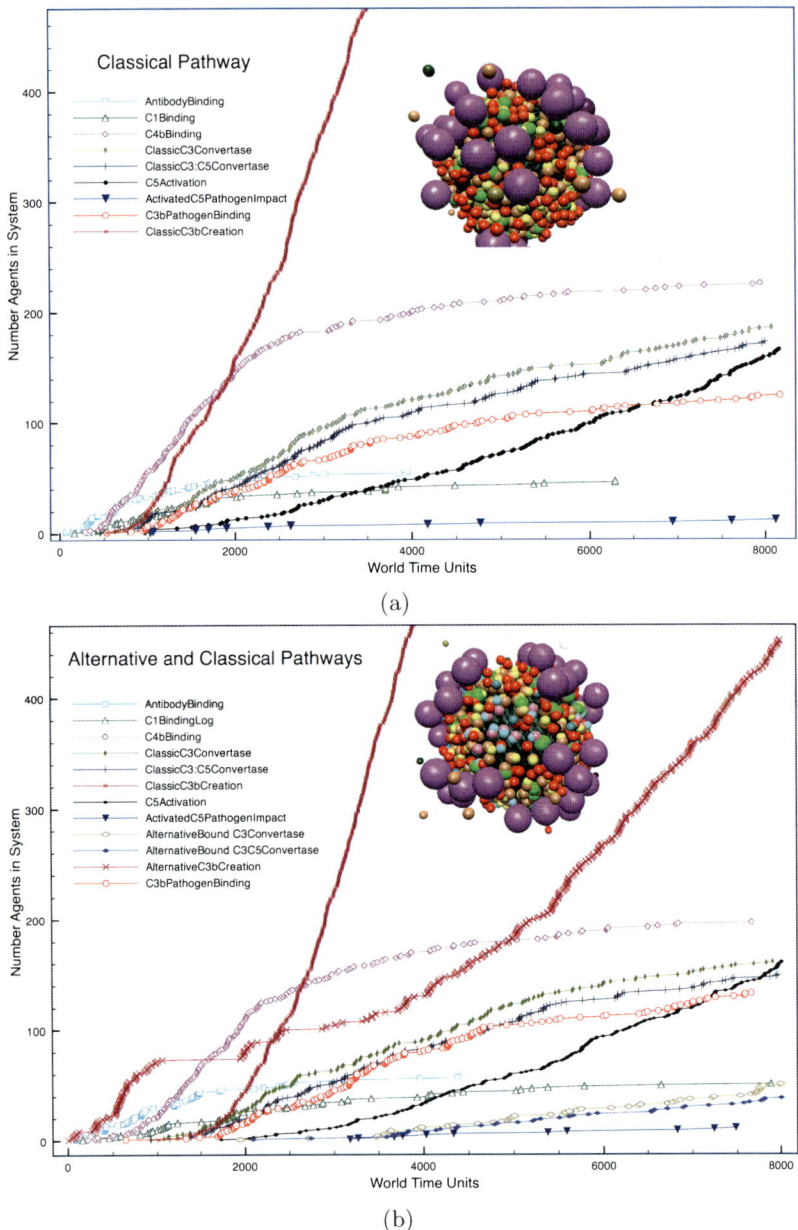

Fig. 23. Agent concentrations as monitored for a typical simulation experiment for the (a) Classical and (b) Alternative and Classical pathways.

It is interesting, though, that under these concentrations the Alternative pathway mechanisms might act as a regulatory process governing the initial C3b production in the Classical pathway. It is also noteworthy that the combination of both pathways results in an increase in overall convertase production. For future experiments, a range of concentrations for all the complement proteins needs to be explored, as each individual person will have their own unique balance and effectiveness of their complement proteome. Furthermore, binding affinities for all the agents in the system need to be integrated into the current model. Other possible modifications would be to include regulatory protein components of the complement system. Rather than (as the current model does) using decay rates on the agents, one would use the regulatory agents to self-regulate the reaction processes.

12 Conclusions

Currently, we only have incorporated some of the earlier and basic theories of how immune system processes might work. Now that we are getting closer to implementing a flexible and powerful simulation infrastructure for biological systems, calibrating and validating our models as well as including more of the recently proposed immune system models [33] is one of our next objectives. We are also expanding our simulations to demonstrate why the generation of effective vaccines is difficult and how spontaneous auto-immunity emerges.

In this contribution, we have tried to give an example of information processing of a highly distributed biological system within the human body: the immune system. In the case of the immune system, information is processed across different time scales—from microseconds for a protein-molecule interaction to days for an immune response to fully unfold. Information processing is also distributed spatially—from the bone marrow, to the spleen and thymus, to lymph nodes, and the lymphatic system, which is tightly interconnected with the blood circulatory system. Hence, the immune system constitutes a prime example of a parallel and highly distributed network of (sub-)systems. Information is carried, memorized and processed in the form of physical shapes, for example, when antigen presenting cells trigger antibody production after successful pattern matching. Memorization of viral signatures through memory B cells is another information processing and storage task that the immune system performs. From a modeling perspective, highly distributed, hierarchical, and networked biological systems provide a rich basis for exploring the effective use of distributed, artificial information processing systems (e.g., computer networks, grid computing, multi-core systems). Studying biological information processing also inspires new methodologies for our future computational engines, with the hope to improve their robustness and flexibility through adaptation, learning and evolution.

Up-to-date details about our latest immune system models and other agent-based simulation examples, which are investigated in our *Evolutionary & Swarm Design Laboratory* can be found at: http://www.swarm-design.org

Acknowledgements

C.J. thanks his former undergraduate students Julius Litorco and Leo Lee for implementing the first prototypes of the simulations, which led to the expanded immune system model discussed in this article.

This work is partly supported by the Undergraduate Medical Education (UME) program at the Faculty of Medicine, University of Calgary as well as grants from the Alberta Heritage Foundation for Medical Research (AHFMR) and from the Natural Sciences and Engineering Research Council (NSERC), Canada.

References

1. Abbas, A.K., Lichtman, A.H.: Basic Immunology - Functions and Disorders of the Immune System. W. B. Saunders Company, Philadelphia (2001)
2. Atamas, S.P.: Self-organization in computer simulated selective systems. BioSystems 39, 143–151 (1996)
3. Bagley, R.J., Farmer, J.D., Kauffman, S.A., Packard, N.H., Perelson, A.S., Stadnyk, I.M.: Modeling adaptive biological systems. BioSystems 23, 113–138 (1989)
4. Bezzi, M., Celada, F., Ruffo, S., Seiden, P.E.: The transition between immune and disease states in a cellular automaton model of clonal immune response. Physica A 245, 145–163 (1997)
5. Biesma, D.H., Hannema, A.J., Van Velzen-Blad, H., Mulder, L., Van Zwieten, R., Kluijt, I., Roos, D.: A family with complement factor d deficiency. J. Clin. Invest. 108(2), 233–240 (2001)
6. Bonabeau, E., Dorigo, M., Theraulaz, G.: Swarm Intelligence: From Natural to Artificial Systems. Santa Fe Institute Studies in the Sciences of Complexity. Oxford University Press, New York (1999)
7. Bower, J.M., Bolouri, H. (eds.): Computational Modeling of Genetic and Biochemical Networks. MIT Press, Cambridge (2001)
8. Burleigh, I., Suen, G., Jacob, C.: Dna in action! a 3d swarm-based model of a gene regulatory system. In: ACAL 2003, First Australian Conference on Artificial Life, Canberra, Australia (2003)
9. Burleigh, I., Suen, G., Jacob, C.: Dna in action! a 3d swarm-based model of a gene regulatory system. In: ACAL 2003, First Australian Conference on Artificial Life, Canberra, Australia (2003)
10. Camazine, S., Deneubourg, J.L., Franks, N.R., Sneyd, J., Theraulaz, G., Bonabeau, E.: Self-Organization in Biological Systems. Princeton Studies in Complexity. Princeton University Press, Princeton (2003)
11. Camazine, S., Deneubourg, J.L., Franks, N.R., Sneyd, J., Theraulaz, G., Bonabeau, E.: Self-Organization in Biological Systems. Princeton Studies in Complexity. Princeton University Press, Princeton (2003)
12. Castiglione, F., Mannella, G., Motta, S., Nicosia, G.: A network of cellular automata for the simulation of the immune system. International Journal of Modern Physics C 10(4), 677–686 (1999)
13. Celada, F., Seiden, P.E.: A computer model of cellular interactions in the immune system. Immunology Today 13(2), 56–62 (1992)

14. Celada, F., Seiden, P.E.: Affinity maturation and hypermutation in a simulation of the humoral immune response. European Journal of Immunology 26, 1350–1358 (1996)
15. Clancy, J.: Basic Concepts in Immunology - A Student's Survival Guide. McGraw-Hill, New York (1998)
16. Farmer, J.D., Packard, N.H.: The immune system, adaptation, and machine learning. Physica D 22, 187–204 (1986)
17. Guo, Z., Han, H.K., Tay, J.C.: Sufficiency verification of hiv-1 pathogenesis based on multi-agent simulation. In: GECCO 2005: Proceedings of the 2005 Conference on Genetic and Evolutionary Computation, pp. 305–312. ACM Press, New York (2005)
18. Guo, Z., Tay, J.C.: A comparative study on modeling strategies for immune system dynamics under HIV-1 infection. In: Jacob, C., Pilat, M., Bentley, P., Timmis, J. (eds.) ICARIS 2005. LNCS, vol. 3627, pp. 220–233. Springer, Heidelberg (2005)
19. Hanegraaff, W.: Simulating the Immune System. Master's thesis, Department of Computational Science, University of Amsterdam, Amsterdam, The Netherlands (September 2001)
20. Hoar, R., Penner, J., Jacob, C.: Transcription and evolution of a virtual bacteria culture. In: IEEE Congress on Evolutionary Computation. IEEE Press, Canberra (2003)
21. Jacob, C., Barbasiewicz, A., Tsui, G.: Swarms and genes: Exploring λ-switch gene regulation through swarm intelligence. In: Congress on Evolutionary Computation. IEEE Press, Vancouver (2006)
22. Jacob, C., Burleigh, I.: Biomolecular swarms: An agent-based model of the lactose operon. Natural Computing 3(4), 361–376 (December 2004)
23. Jacob, C., Litorco, J., Lee, L.: Immunity through swarms: Agent-based simulations of the human immune system. In: Nicosia, G., Cutello, V., Bentley, P.J., Timmis, J. (eds.) ICARIS 2004. LNCS, vol. 3239, pp. 400–412. Springer, Heidelberg (2004), http://www.springeronline.com/sgw/cda/frontpage/0,11855,4-40109-22-34448379-0,00.html
24. Jacob, C., von Mammen, S.: Swarm grammars: growing dynamic structures in 3d agent spaces. Digital Creativity 18(1), 54–64 (2007), http://www.tandf.co.uk/journals/ndcr
25. Jacob, C., Steil, S., Bergmann, K.: The swarming body: Simulating the decentralized defenses of immunity. In: Bersini, H., Carneiro, J. (eds.) ICARIS 2006. LNCS, vol. 4163, pp. 52–65. Springer, Heidelberg (2006)
26. Janeway, C.A., Travers, P., Walport, M., Shlomchik, M.J.: Immunobiology: The Immune System in Health and Disease, 6th edn. Garland Science, New York (2005), www.garlandscience.com
27. Johnson, S.: Emergence: The Connected Lives of Ants, Brains, Cities, and Software. Scribner, New York (2001)
28. Kleinstein, S.H., Seiden, P.E.: Simulating the immune system. Computing in Science & Engineering 2(4), 69–77 (2000)
29. Kuby, J., Goldsby, R.A., Kindt, T.J., Osborne, B.A.: Immunology, 6th edn. W. H. Freeman and Company, New York (2003)
30. von Mammen, S., Jacob, C.: Genetic swarm grammar programming: Ecological breeding like a gardener. In: IEEE Congress on Evolutionary Computation, Singapore (September 25-28, 2007)
31. von Mammen, S., Jacob, C.: The spatiality of swarms. In: Artificial Life XI, 11th International Conference on the Simulation and Synthesis of Living Systems, Winchester, UK (2008)

32. von Mammen, S., Jacob, C.: The evolution of swarm grammars: Growing trees, crafting art and bottom-up design. IEEE Computational Intelligence Magazine (2008) (under review)
33. Meier-Schellersheim, M., Fraser, I., Klauschen, F.: Multi-scale modeling in cell biology. Wiley Interdiscip. Rev. Syst. Biol. Med. 1(1), 4–14 (2009)
34. Muller-Eberhard, H.J.: Complement. Annu. Rev. Biochem. 44, 697–724 (1975)
35. Noble, D.: The Music of Life. Oxford University Press, Oxford (2006), http://www.oup.com
36. Nossal, G.J.: Life, death and the immune system. Scientific American, 53–62 (September 1993)
37. Parham, P.: The Immune System. Garland Publishing, New York (2000); in my home library since 2003
38. Penner, J., Hoar, R., Jacob, C.: Bacterial chemotaxis in silico. In: ACAL 2003, First Australian Conference on Artificial Life, Canberra, Australia (2003)
39. Pogson, M., Smallwood, R., Qwarnstrom, E., Holcombe, M.: Formal agent-based modelling of intracellular chemical interactions. BioSystems (2006)
40. Przybyla, D., Miller, K., Pegah, M.: A holistic approach to high-performance computing: xgrid experience. In: ACM (ed.) Proceedings of the 32nd annual ACM SIGUCCS conference on User services, pp. 119–124 (2004)
41. Puzone, R., Kohler, B., Seiden, P., Celada, F.: Immsim, a flexible model for in machina experiments on immune system responses. Future Generation Computer Systems 18(7), 961–972 (2002)
42. Rössler, O., Lutz, R.: A decomposable continuous immune network. BioSystems 11, 281–285 (1979)
43. Salzberg, S., Searls, D., Kasif, S. (eds.): Computational Methods in Molecular Biology, New Comprehensive Biochemistry, vol. 32. Elsevier, Amsterdam (1998)
44. Smith, L.C.: The ancestral complement system in sea urchins. Immunological Reviews 180, 16–34 (2001)
45. Sompayrac, L.: How the Immune System Works, 2nd edn. Blackwell Publishing, Malden (2003)
46. Spector, L., Klein, J., Perry, C., Feinstein, M.: Emergence of collective behavior in evolving populations of flying agents. In: Cantú-Paz, E., Foster, J.A., Deb, K., Davis, L., Roy, R., O'Reilly, U.-M., Beyer, H.-G., Kendall, G., Wilson, S.W., Harman, M., Wegener, J., Dasgupta, D., Potter, M.A., Schultz, A., Dowsland, K.A., Jonoska, N., Miller, J., Standish, R.K. (eds.) GECCO 2003. LNCS, vol. 2723, pp. 61–73. Springer, Heidelberg (2003)
47. Tarakanov, A., Dasgupta, D.: A formal model of an artificial immune system. BioSystems 55, 151–158 (2000)
48. Tay, J.C., Jhavar, A.: Cafiss: a complex adaptive framework for immune system simulation. In: SAC 2005: Proceedings of the 2005 ACM Symposium on Applied Computing, pp. 158–164. ACM Press, New York (2005)
49. Walport, M.J.: Complement. first of two parts. N. Engl. J. Med. 344, 1058–1066 (2001)
50. Wolfram, S.: A New Kind of Science. Wolfram Media, Champaign (2002)

Biological Limits of Hand Preference Learning Hiding Behind the Genes

Fred G. Biddle and Brenda A. Eales

Department of Medical Genetics, Department of Biochemistry and Molecular Biology,
and Department of Biological Sciences, University of Calgary, Calgary, Alberta, T2N 4N1, Canada
fgbiddle@ucalgary.ca

Abstract. Biological limits of learning are the cto our understanding of learning and memory. We are conducting a detailed investigation of structure and function of gene regulatory networks in a biological system of learning and memory. We use a simple system of hand reaching in laboratory mice and assess the regulation of the limit of hand preference learning. In this chapter, we describe hand-reaching behavior of mice and the discovery that it is a complex adaptive behavior in which future preference is genotypically dependent on past experience. We define the key elements that led to the collaborative development of a stochastic agent-based model. Simulation with the model mimics hand-reaching behavior and successfully predicts dynamics and kinetics of the learning and memory process in different genotypes of mice. Therefore, a mouse receives information from the act of reaching and uses it to inform the choice of the next hand reach; the model shows the structure of the behavior and the biological limits of hand preference learning hiding behind the genes. We use three different mouse strains as prototypes to illustrate the constructive path that predicts the adaptive behavioral phenotype from specified genotype. We believe that promises made by the power of genetic analysis will be kept by the discovery that simplicity resides in gene regulatory networks that give rise to learning and memory in adaptive behaviors.

"The 'structural' concept of gene action accounts for the multiplicity and for the phylogenetic stability of macromolecular structure. It does not account for biochemical coordination and ignores the problem of emergence and functioning of differentiated cellular populations" [1].

1 Introduction

Asymmetry of human hand preference continues to attract creative speculation about its biological cause and the evolutionary significance of its phenotypic diversity. However, research on laterality of our handedness has mostly added to a rich mythology about the behavior, including its unavoidable association with asymmetry of brain development and function [2,3,4,5,6]. Family analysis and twin studies in different populations continue to support a heritable influence on asymmetry of hand preference [7] and the role of genes remains central to speculation about its biological cause and evolutionary significance [8,9]. The concern is that relevant genes are not yet known, despite our ability to sequence and analyze the genome [10]. It may be that analysis of hand preference has been trying to detect genetic effects one gene at a time, similar to what is done in some experimental model systems with directed mutation and gene knock-out technology [11]; if many genes influence the behavior, but they have complex effect and only a small effect size, they may be virtually impossible to detect one gene at a time. Therefore, we suggest that genetic research on human hand preference may be trying to attach the wrong model of its behavior to the genome. Biology of hand preference may be at the same impasse in the analysis of genes and behavior that Jacob and Monod [1] identified 50 years ago for molecular biology where the structural concept of gene action accounted for "multiplicity ... and stability of macromolecular structure", but the concept of genes did not reveal "biochemical coordination" and it ignored "emergence and functioning of differentiated cellular populations." Biochemical coordination and the emergence and functioning of differentiated cellular populations were only understood when the science of metabolism was separated from the science of code and control.

We work on the genetic analysis of mouse models of complex traits of mammalian development. We believe that genetically regulated experimental models can provide a rigorous framework to objectively assess and understand complex traits of human development. Laboratory mice have an extensive array of different, genetically uniform and well-characterized inbred strains and mutant-carrying stocks [12] and a systematic analysis of the wide range of diversity in this genetic resource can reveal the biologically possible range of phenotypic diversity in complex traits of mice [13,14]. An inbred strain of laboratory mice provides the key element of reproducible experiments within a set of genetically identical individuals and the collection of inbred strains provides the ability to compare the experiments between genetically different individuals. Furthermore, mice have asymmetry in their paw preference in single-paw reaching tasks and we demonstrated that right- and left-paw preference is a complex adaptive behavior that reflects a dynamic biological system of learning and memory[15,16,17,18]. Research on the structural elements of this mouse behavior provides a practical experimental paradigm to investigate the biology of an adaptive behavior and the evolutionary significance of its heritable variation. The mouse behavior is giving some insight into human hand-preference behavior.

In this review, we describe the paw preference of mice and briefly summarize early investigation of its phenotypic diversity among different mouse strains [19,20]. Discovery of the system of learning and memory in the mouse behavior was an unexpected surprise [15]. Initial structural analysis of the system [16] led to an agent-based stochastic model that replicates the complexity in the behavior [17]. Simulations with the model reproduced the dynamic variation in paw-preference learning of different mouse strains at both the individual and population levels and they predicted the limits of learning. The model confirmed that paw preference is an adaptive behavior. New discoveries with the mouse model suggest new experiments to assess the biological significance of diversity in paw-preference behavior and the rationale for these new experiments derives from the hypothesis that the behavior is the expression of a genetically regulated, learning and memory process [18].

Our new experiments assess the biological information in paw-preference behavior. They show that an individual mouse receives information from each paw reach, uses it to influence the choice of the next reach and, hence, establishes a reliable paw preference. We describe some of the experiments that are in progress because they define critical genetic questions that may form a foundation for practical investigations about the evolutionary significance of the mouse behavior. The experiments confirm new predictions from our hypothesis that the mouse behavior is a genetically regulated process of learning and memory and they provide a realistic framework to objectively assess human hand preference.

Geneticists have two distinctive uses for the word gene [21]. One is a unit of inheritance or code that allows a trait to be reliably transmitted to the next generation (either the two daughter cells of a parent cell or the offspring of specific individuals); the other is a unit of control that specifies molecular elements that function and interact in the cell or individual. Therefore, if heritable differences influence paw preference of individual mice, the question is what is transmitted from generation to generation so that parents, offspring, grand offspring, etc., have a predicted paw preference? What do genes do so that an individual expresses the predicted preference and will the 'study of genes' tell us anything about the behavior? We suggest that, in order to answer these questions, the science of paw-preference behavior must be separated form the science of code and control because mice reach and express a reliable paw preference, genes dont. Therefore, we ask what is the paw-preference behavior that is hiding behind the genes?

2 Mouse Paw Preference

2.1 Paw Preference Phenotypes Are Characteristic Behavior Patterns

After a brief fasting from food, mice will reach to retrieve food from a tube in a small testing chamber (Fig. 1). When the food tube is centered in the face of the test chamber, it is defined as an unbiased or U-world and a mouse is free to reach with its right and left fore paw. When previously untested mice are allowed a set number of paw reaches in the U-world, they will express a reliable ratio of right and left paw reaches. The right-paw entry or RPE score quantifies the direction of paw preference and it is the number of reaches made with the right paw in a specified number of reaches [19]. For example, if mice are allowed 50 paw reaches, the RPE score will have a numerical value from 0 to 50. Extensive surveys showed that most inbred strains have equal numbers of right and left-paw mice in a U-world [19,22,23,24,25,26]. Figure 2 illustrates the characteristic and reliable patterns of U-world RPE scores from four genetically different mouse strains. The individual mice of an inbred strain are genetically identical because inbreeding, usually by sister brother matings, within the strain for hundreds of generations has eliminated genetic variation by chance segregation.

Fig. 1. Mouse reaching for food with right forepaw in unbiased U-world [24]. In U-world, food tube is centered in face of test chamber; in biased R-world or L-world, food tube is directly next to the respective right or left side from the perspective of the mouse. (Reproduced with permission from NRC Research Press. (c) 2008 NRC Canada or its licensors.)

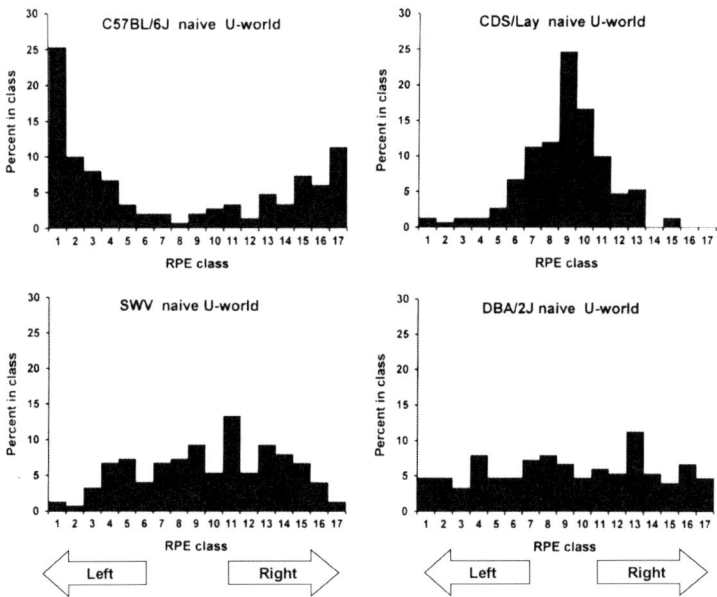

Fig. 2. Diversity in dynamic patterns of paw preference in the U-world from C57BL/6J, CDS/Lay, SWV, and DBA/2J mouse strains [41]. Sample size is 150 mice with equal number of females and males from each strain. Previously untested mice were allowed 50 paw reaches to retrieve food from the food tube; number of right-paw reaches (RPE score) is the measure of direction of paw preference; the 51 RPE scores (0 50 reaches with the right paw) are binned in 17 equal sized classes of 3. (Reproduced with permission from NRC Research Press. (c) 2008 NRC Canada or its licensors.)

The characteristic pattern of U-world RPE scores is the heritable property of a specific strain since it is impossible to select a right or left direction of paw preference within an existing inbred strain [20]. Genetic differences between mouse strains must give rise to the reliable patterns of RPE scores (Fig. 2) and those heritable differences must cause a difference in degree of lateralization of the preferred paw of individual mice rather than a difference in the right or left direction of their preferred paw. Selective breeding studies with hybrid mice, derived from matings between different inbred strains, confirmed that degree of lateralization of the preferred paw is the heritable mouse trait [27,28]. The genes and their function remain unknown in the sense of code and control. They are assumed to be complex.

2.2 Paw Preference and Other Developmental Traits

If a behavioral trait is complex, more than one gene may be involved and the functional interactions among genes may also be complex. Therefore, alternate methods of analysis, such as potential brain or neural correlates of the behavior, might provide insights to the behavior rather than a direct attempt to identify and isolate genes [29]. Many associations have been identified between different developmental traits and differences in lateralization of mouse paw preference. The associations have added layers of complexity without detecting the biological cause of the diversity in the behavior. For example, there may be a dependence of the behavior on size (including absence) of the corpus callosum (a major inter-cerebral hemispheric neuronal fiber tract) in some strains but not in others [26,30,31,32]. Similarly, there may be a relation of lateralization to large differences between strains in the size of the hippocampal neuronal projections [33]. Variations in different measures of immune reactivity have an association with the behavior in some but not other strains [34,35,36,27].

2.3 Genes and Degree of Lateralization of the Preferred Paw

The patterns of paw preference scores from the U-world tests of different inbred strains (Fig. 2) suggested that the heritable trait is the degree of lateralization of the preferred paw, without regard to its left or right direction [19,20]. In Figure 2, C57BL/6J is highly lateralized in preferred paw use and CDS/Lay is weakly lateralized or ambilateral; the SWV and DBA/2J strains are intermediate in degree lateralization. Since 50 paw reaches were used, one measure of degree of lateralization of the preferred paw was simply to fold the RPE distribution at the midpoint of 25 and redefine the score as a preferred paw entry (PPE) for each mouse. Using degree of lateralization as the measure of paw preference, both highly lateralized and weakly lateralized true breeding lines were rapidly selected from a genetically heterogeneous hybrid stock of mice that was constructed from the matings of eight different mouse strains [27,28]. The recovery of true breeding alternate phenotypes in selected generations was the genetic evidence that degree of lateralization is a heritable trait, but it did not identify number and function of the responsible genes. A quantitative trait locus (QTL) breeding analysis, using two mouse strains that differed significantly in degree of lateralization of their preferred paw, detected positive association with marker genes on mouse chromosome 4 [38]; however, no genetic test of transmission has confirmed that chromosome 4 genes can predict differences in degree of lateralization. Also, we applied a QTL analysis to the difference in degree of lateralization between the C57BL/6J and CDS/Lay strains (see Fig. 2) and failed to detect any evidence for segregation of genetic heterogeneity in the trait [15]. Therefore, we looked in other ways at the patterns of paw preference behavior.

2.4 Biased Test Chambers Demonstrate Learning and Memory

In the unbiased U-world (Fig. 1), the mouse is allowed to choose to reach with its right- or left-paw. Instead, we used an alternate testing chamber, previously developed by Collins [39], in which the food tube is placed flush to the right side (R-world) or left side (L-world) from the perspective of the mouse. It is presumably more difficult for a mouse to reach in a biased test world with the paw that is opposite to the world bias and we found greater phenotypic variation in biased test worlds than was visible in the U-world [15,40,41]. Previously untested C57BL/6J mice expressed dramatically different patterns of paw preference in the biased test chambers when compared to the behavior of previously untested CDS/Lay mice (Fig. 3). There is gene-environment interaction between the average RPE score and the direction of the test chamber (Fig. 4); C57BL/6J mice are more left handed on average than CDS/Lay mice in a L-world, but CDS/Lay mice are more left handed than C57BL/6J mice in a R-world. The direction of paw preference changed in response to the direction of the biased world and the

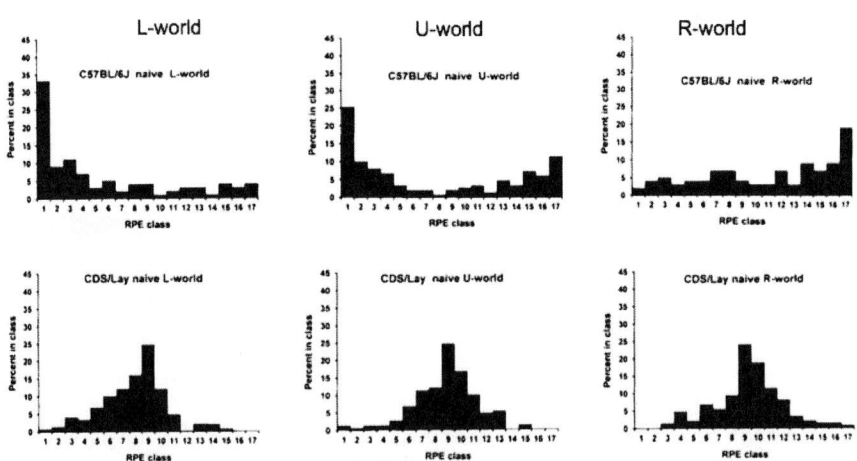

Fig. 3. Dynamic patterns of right-paw entry (RPE) scores from previously untested C57BL/6J and CDS/Lay mice in biased L-world and R-world compared with patterns in unbiased U-world [15]. Sample size is 150 mice and equal number of females and males from each strain. RPE scores are binned as in Fig. 2. (Reproduced with permission from NRC Research Press. (c)2008 NRC Canada or its licensors.)

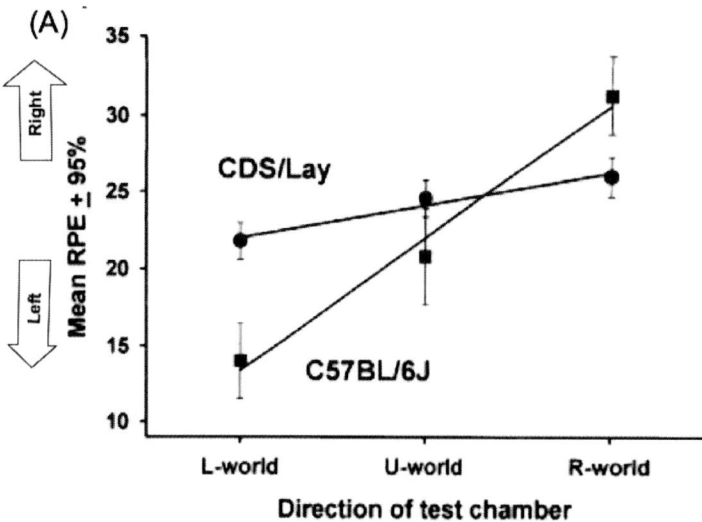

Fig. 4. Gene - environment interaction in direction of paw preference expressed by previously untested C57BL/6J and CDS/Lay mice in response to direction of test world [15]. Mean (average) RPE and 985% confidence limits are from Fig. 3. Average direction of C57BL/6J is more left-handed than CDS/Lay in a L-world, but it is more right-handed than CDS/Lay in a R-world. (Modified and reproduced with permission from NRC Research Press. (c) 2008 NRC Canada or its licensors.)

relative order of the difference in direction between the strains changed in the oppositely biased worlds. More importantly, the directions of paw preference of the tested mice did not return to their expected baselines when the mice were reassessed one-week later in the oppositely biased test worlds. Therefore, the experience of reaching for food had conditioned the behavior and the mice had learned a preference in the context of their first biased-world test.

2.5 Kinetics Uncovered Four Elements of Paw Preference Learning and Memory

A detailed kinetic analysis revealed four key elements that involve time in the learning and memory process of paw preference behavior [16]. The analysis was done by training C57BL/6J mice in a L-world and then assessing their paw preference in a R-world. Selected measures of the learning process were compared between other strains and hybrids and the highlights are noted here.

Memory of 50 training reaches in a L-world was consolidated over time after training (Fig. 5). When independent groups of mice were retested in the R-world, direction of paw preference changed from being right-pawed in a R-world to being left-pawed in response to elapsed time after L-world training. Kinetic analysis demonstrated that contextual memory of L-world training would asymptotically reach a maximum, but half the estimated amount was achieved in 1.4 days.

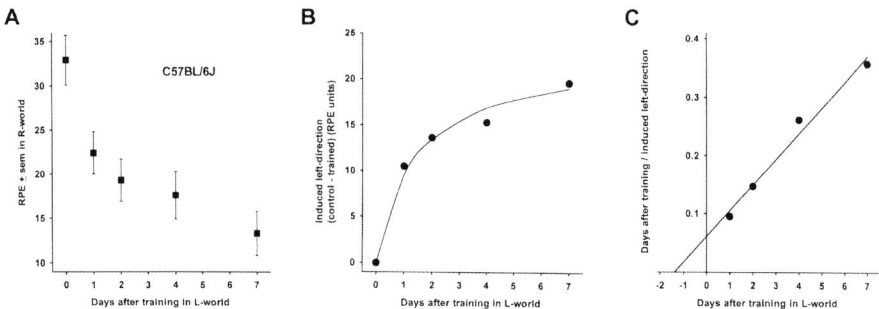

Fig. 5. Consolidation of paw preference memory in C57BL/6J mice with time after training in L-world with 50 reaches [16]. Direction of hand preference (RPE) is measured in opposite R-world at different times after L-world training. (A) RPE in R-world becomes more left-handed with time after training. (B) Relative amount of change in preference asymptotically approaches a maximum. (C) Half-maximum consolidation is estimated from inverse of slope of linear transformation; time to achieve half-maximum consolidation is estimated from intercept with x-axis. (Reproduced with permission from NRC Research Press. (c)2008 NRC Canada or its licensors.)

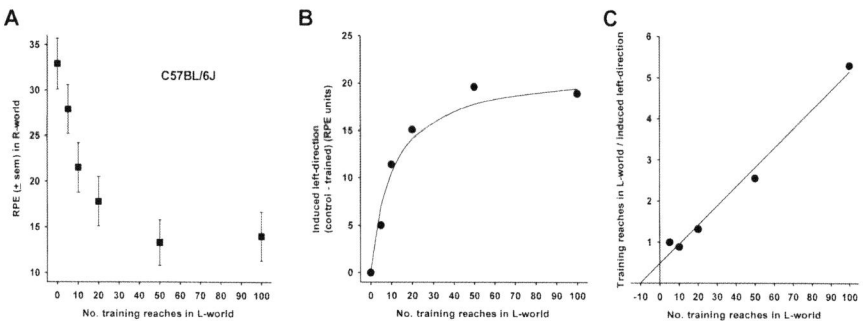

Fig. 6. Kinetics of memory acquisition in response to training reaches in C57BL/6J [16]. One week after different numbers of training reaches in L-world, the mice were allowed 50 reaches in the R-world. (A) Hand preference in R-world becomes more left-handed in response to more L-world training reaches. (B) Relative direction of preference asymptotically approaches a maximum. (C) Half-maximum capacity of memory is estimated from inverse of slope of linear transformation; ability or number of training reaches to achieve the half-maximum capacity if estimated from intercept with x-axis. (Reproduced with permission from NRC Research Press. (c)2008 NRC Canada or its licensors.)

Administration of the protein-synthesis inhibitor, anisomycin, immediately after training blocked the consolidation of the contextual memory of paw preference training in a dose-response manner. This implied that gene activity and new protein synthesis are necessary for this learning and memory process.

If memory of training required time to consolidate, the memory of the L-world training also decayed over time. Decay of memory was exponential and the remaining memory was lost at a constant rate with an estimated half-life of 6.4 weeks. Those results prompted us to remark that the interpretation the paw preference behavior is critically dependent on prior experience of the mice. For example, if 6.4 weeks is the half-life of the memory of 50 training reaches, approximately 32 weeks (or 5 half-lives) would have to elapse in order to lose greater than 95.

The speed of learning or rate of learning in response to different numbers of training reaches was estimated by training independent groups of C57BL/6J mice. Again, the mice were training in the L-world and they were retested one-week later in the R-world (Fig. 6). The direction of paw preference in the R-world changed from being right-pawed to being left-pawed in response to the L-world training reaches and it asymptotically approached a point where more training had no further effect. C57BL/6J mice achieved their estimated half-maximum capacity to learn a direction of preference with only 10.4 L-world training reaches (Fig. 6) and this was significantly faster than other strains and hybrids that were tested [16].

3 Stochastic Agent-Based Model of Mouse Paw Preference

Paw preference of mice is clearly an adaptive behavior that depends on genotype and environment and prior experience. At the population level, paw preference is a probabilistic behavior and genetically identical mice exhibit individual-to-individual diversity in their paw preference when they are subjected to identical experimental tests (Fig. 2). Different strains of mice have characteristically different distributions of behavioral response. Measurements of gene expression and different biochemical processes in other contexts have demonstrated that the dynamics in biological systems is stochastic [42,43] and different vital processes can have a probabilistic nature [44,45]. Noise in gene expression generates phenotypic variability [46]. Therefore, critical information about regulatory mechanisms of mouse paw preference behavior may be contained in the reliable but noisy patterns of paw preference. To test this possibility, a stochastic agent-based model was developed to reproduce paw preference behavior at the individual and population levels [17]. If the model were successful in predicting the behavior, it would be useful to relate the noisy paw preference phenotype to the deterministic genotype of different mouse strains.

3.1 Stochastic Model Mimics Mouse Paw Preference

A surprisingly simple, stochastic agent-based model was constructed to mimic the paw reaching events of single mice [17]. At each consecutive time interval in the model, the right or left paw reach of a model mouse is assessed for a match to the right or left paw that the mouse ought to use to reach for food in a test-world

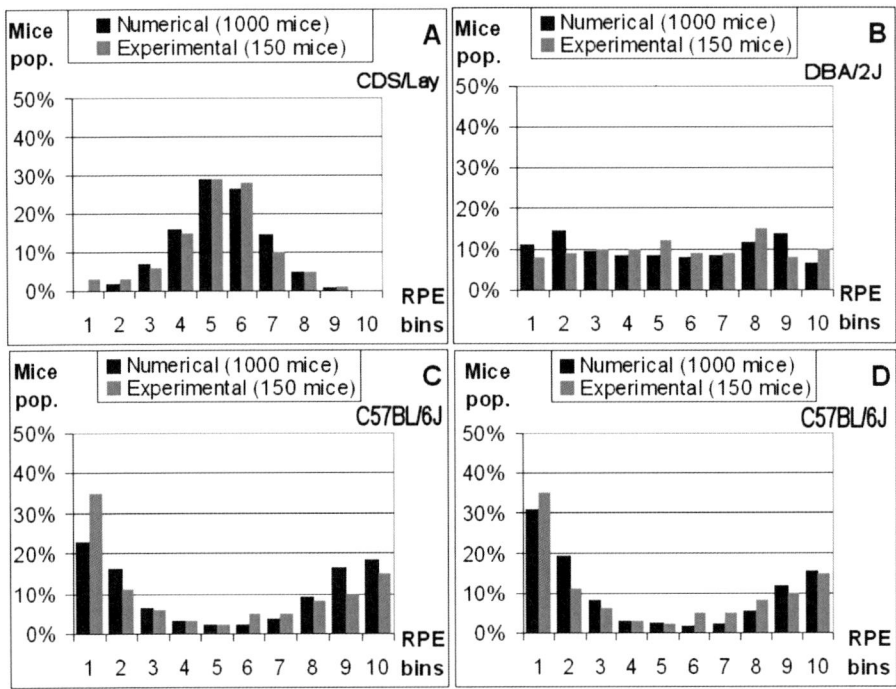

Fig. 7. Numerically simulated model mice match genetic diversity in dynamic patterns of U-world hand preference of experimental mice [17]. Naive mice were allowed 50 reaches and RPE scores are binned in 10 bins as a percent of the population. (A) Simulated CDS/Lay model mice with a training rate of 25 memory units matched the pattern of RPE from experimental mice. (B) Simulated DBA/2J mice with a training rate of 60 memory units matched experimental mice. (C) Simulated C57BL/6J mice did not match experimental with a training rate of 95 memory units, but (D) matched experimental mice with a biased training rate of 105 left-hand memory units and 85 right-hand memory units. (Reproduced by Permission of SAGE.)

that has a probabilistically defined bias. A model mouse receives a number of right- or left-memory units from a fixed number of memory units for the correct matching of paw reach with test world and, in order to maintain a dynamic equilibrium, the model loses memory at a constant rate for both a successfully matched and an unsuccessfully matched reach. Probability of using either the right or left paw changes during a simulation by two processes: (1) each time a mouse reaches, it records a successful reach and increases the probability of using that paw in the next reach; (2) the amount by which the probability varies corresponds to how much the mouse learns from a successful reach. Training (learning) rate is the deterministic genetic trait of each mouse strain. The model also allowed the amount of learning with the right paw to differ from the left paw if that bias was found to be necessary.

The model was tuned in the U-world with different training rates. Numerical mice were given 50 training reaches and 1000 mice were numerically simulated and compared with 150 experimental mice from the CDS/Lay, DBA/2J and C57BL/6J strains (Fig. 7). The simulated left-paw and right-paw training rates of 25 memory units and 60 memory units matched the experimental results for CDS/Lay and DBA/2J, respectively, but a left-bias of 105 left-memory units versus 85 right-memory units was necessary to match the experimental results from C57BL/6J mice. The relationship between training rate and decay of memory shapes the distribution of RPE paw-preference scores. Further analyses with different numbers of training reaches and different training rates showed that the RPE distributions go from Gaussian-like to flat to bimodal U-shape (Fig. 7). Simulations demonstrate that genetically identical mice give rise to a distribution of RPE scores rather than a single distinct RPE score. Therefore, RPE scores of biological mice are the result of the stochastic nature of paw-reaching behavior and the interactions between each mouse and its environment.

The immediate value of simulations with the agent-based model was their demonstration of a detectable amount of paw-preference learning in all mice. Previously, we had described the paw preference in some strains, such as CDS/Lay and DBA/2J, as a constitutive behavior because their learning rate was below a level that we could detect by standard null hypothesis significance testing procedures [15,16,40,41]. Now, we consider simulations with the agent-based model to be an essential tool because they provide the two crucial pieces of biological information of (1) magnitude of the effect of interest and (2) precision of the estimate of the magnitude of that effect [47,48]. We illustrate that fact with estimation of the limits of learning.

3.2 Stochastic Model Predicts Limits of Learning

Different training rates allowed the stochastic model to match the U-world behavior of genetically different experimental mice. Validity of the model was demonstrated when simulations matched biological observations of replicate testing of mice in a U-world with a one-week interval between tests without any further need to tune the model [17]. A more critical test of the stochastic model is that it predicted the limits of learning in biased test worlds, again without any need to further tune the model. Kinetic analysis had previously demonstrated a limit of learning occurs in C57BL/6J (Fig. 6) as the change to left-hand preference in response to L-world training approaches a saturation point, where more training in the L-world has no measurable effect on direction of hand preference in the R-world. In this case, decay of memory during the one-week interval plays an important role to establish the limit. We simulated 100 independent populations of C57BL/6J model mice with the different numbers of L-world training reaches and retested the populations with 50 reaches in the R-world (Fig. 8). The averaged results of the model mice, along with their standard deviations, matched the biological results almost perfectly.

Numerical simulations experimentally assessed the stochastic nature of the agent-based model of paw preference. Once the training rate of the model was

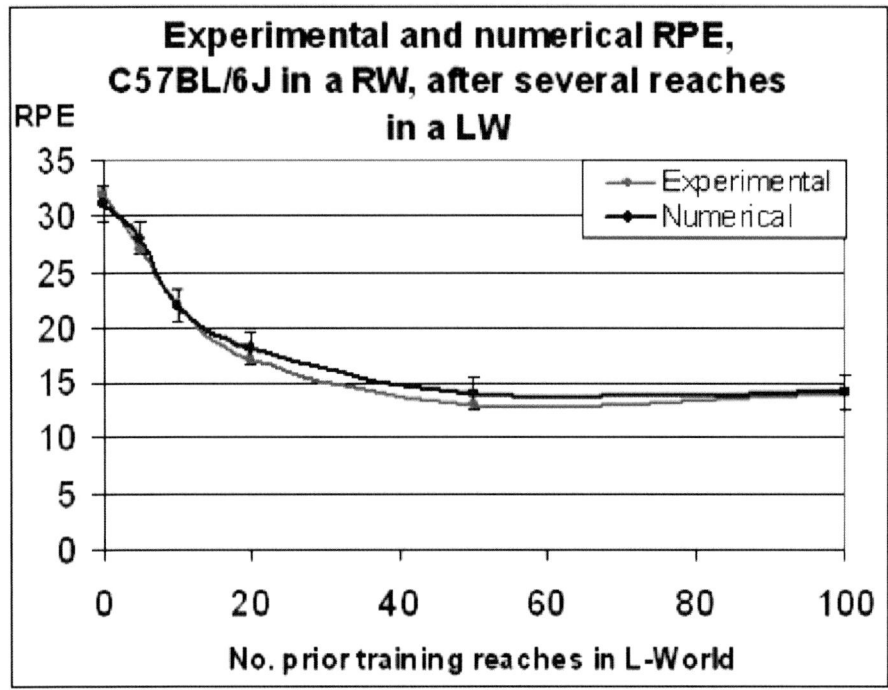

Fig. 8. Numerically simulated model mice match kinetics and limit of learning ability in C57BL/6J experimental mice [17]. Numerical and experimental mice received 50 reaches in R-world one week after training with different numbers of L-world training reaches; means and standard deviations are from 100 simulated populations. (Reproduced by Permission of SAGE.)

tuned for each genotype, further simulations predicted the paw-preference behavior in different environments. The model predicted the future behavior of previously tested mice; hence, the dynamics of relearning must be identical to the dynamics of first-time learning in naive mice. Since memory of prior training also decays in a simple way, there is no detectable residual effect of training on the dynamics of relearning paw preference the mouse does it again with the same rate and the same limits.

3.3 Simulations Suggest New Experiments and Identify a Short-Term Memory

Success with the agent-based model suggests new experiments to assess the learning and memory process of paw preference behavior [18]. Several examples are briefly outlined. Mice obviously learn a direction of paw preference in response to reaching in biased worlds where we expected that it is more difficult to reach for food with the paw that is opposite to the direction of the bias [18].

Nevertheless, it is not clear why a mouse should learn a preference in the unbiased U-world where reaching with either paw provides equal reward and we assume that it involves the same effort. Also, if a mouse learns a paw preference in the U-world, does preference change during the training session or is it detectable only between training sessions with a period for memory consolidation? Do mice have a constitutive bias in paw preference and is it important relative to an acquired preference?

Ability to simulate and assess the reach-to-reach behavior of model mice provides a method to evaluate the U-world behavior of individual mice with different training rates. Model nonlearners expressed the reaching behavior and numerical measures of paw preference that we expect in individuals with random paw choices [18]. New measures demonstrated that mice with genetically different training rates learn a paw preference within a U-world training session as well as between training sessions. Paw preference is probabilistic, but a positive autocorrelation between paw choices made by individual mice showed that paw preference changes gradually from reach to reach within a training session and it is concordant between training sessions. The results also suggested that strong biases in paw preference originate by chance from initially weak biases and increase by a positive feedback mechanism during training. Therefore, individual mice can become strongly lateralized in a left or right direction when no bias would be evident from the population average preference score (the approximate average RPE score is 25 for each strain in Fig. 2). Decrease in positive autocorrelation with lag between paw choices demonstrated that a constitutive paw preference plays only a minor role and adaptation to environmental change is the main cause of mouse paw preference. Interestingly, positive autocorrelation in strong learners decreases faster with increasing lag between reaches than it does in weak learners. That observation supports the hypothesis that learning rate and the resulting degree of learning is the deterministic genetic trait. It also raises new questions about the elements of short-term memory that appear in the behavior during a training session.

4 Biological Limits of Paw Preference Learning

4.1 Phenotype and Genotype Versus the Reductive - Constructive Paradigm

Characterizing phenotypes and phenotypic alternates between individuals has been the pathway to gene discovery for molecular biology (Fig. 9). At the level of protein polymorphisms, the forward and reverse path from DNA (gene) to RNA and protein is central dogma; phenotype reveals 'code' through classic reductive analysis based on the gene as the unit of inheritance and the gene product from this code provides a constructive path to phenotype, usually in a seamless, almost one-step process. This paradigm is also the foundation for classical positional cloning and identification of new genes in the genome. It is based on phenotype (and phenotypic alternates) that can be mapped or associated with other genes that have been previously mapped and physically placed

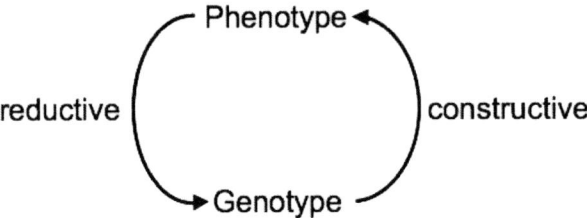

Fig. 9. Reductive and constructive directions of the phenotype genotype relationship.

on the molecular genome and, hence, those other genes serve as cumulative markers of the 'roadmap' of the genome. Mapped location from the study of genetic transmission of a new trait through several generations in informative families reduces phenotype to molecular DNA sequences (code) in the chromosomal 'neighborhood' of the marker genes. The belief is that novel variants in the known molecular sequence of nucleotides in the genomic neighborhood will identify the new molecular genes and functions and the genomic nucleotide sequence will construct the specific phenotypes in the reverse direction.

The reductive-constructive process has been straightforward for simple heritable traits with 'genotypes' that have fully penetrant alternate phenotypes. Classic human metabolic disorders of cystic fibrosis and phenylketonuria (PKU) are good examples of these simple traits. Nevertheless, deeper study of complexity in the phenotypic heterogeneity in the simple traits like PKU is revealing an enormous and unexpected heterogeneity in both genetic code and functional interactions of the code with genetic context and environmental background [49,50]. The apparent single-gene metabolic traits focused our attention to a deeper understanding of cellular processes, but they deflected our attention away from the biology of the whole system of the individual that brought the original phenotype to our attention. For classic PKU, deficiency of the enzyme and resulting higher plasma levels of its substrate, phenylalanine, has not given any functional understanding of the biology of the associated mental retardation. In parallel, many common human diseases are thought to have a common genetic cause in different populations, but the less penetrant phenotypes of the genes are assumed to result from interactions with environment (such as age, diet, toxins, etc.). Some examples of such diseases, again from clinical genetics, are the acute coronary syndromes that are amenable to reductive analysis by population genetic screening with genomic marker genes. They reveal a plethora of suggested genomic coding regions, but in a constructive direction (Fig. 9), almost none of the putative genes have predicted an acceptable degree of significant risk for disease expression through an impaired molecular function [51]. Therefore, knowing the gene in an individual and a statistical risk of disease is not helpful to predict that the individual will actually express the disease. Therefore, how much do we need to know about genes is a sobering societal question because 'risk of disease' is not a disease [52].

Low predictability of risk from reductive-constructive studies of complex human traits emphasizes the need for new approaches in the application of genetics and genomics to biological causation. What is a 'gene' is less clear in a functional sense if several different genomic regions in a coding sense are involved and they interact in different nonlinear ways to influence the expression of phenotype. Complex genetics shows that genetic traits are not contained in individual genes; rather they are properties generated by dynamic networks. 'Code' lies in the interactions among genes and gene products and an observed phenotype may be further contingent on nonlinear change in these interactions in different environments. For complex human traits, 'more is different' in the same sense of different that was emphasized by Philip Anderson [53]. Recognition and resolution of complexity is the challenge of complex traits.

4.2 Paw Preference Biology 'Lives on the Line'

So far, no gene has predicted an individuals behavioral phenotype in the constructive direction, starting from genomic information (Fig. 9). Genomic marker genes have been associated with hand preference behavior, but no gene has predicted the hand preference of an individual human or mouse [38,54,55]. However, for mouse paw preference behaviour, we are near the point where we can write the equation that specifies the behavior of a single genotype and the difference in behavior between different genotypes. We describe this with the phrase biology lives on the line. Paw reaching is stochastic and paw preference is an adaptive behavior in which future behavior depends on past experience. Simulations with agent-based models predict the behavior of individual mice; individual model mice with a specific learning rate can be followed from reach to reach and populations of model mice match the patterns of paw preference behavior that are expressed by biological mice [17,18]. Therefore, if the genotype and its associated training rate can be specified, an individuals paw preference can be probabilistically predicted and the behavior of the genotype under different environmental conditions can be exactly specified. Knowledge of learning rate predicts genotypic limits of learning. Therefore, if we can write the equation that specifies the behavior of a single genotype, we are at least at the point where we can describe the mouse phenotype that is hiding behind the genes.

Improvement ratio (IR) and weighted improvement ratio (WIR) are measures of adaptability of paw preference. They are population measures, but each strain is a population of a single genotype. Mice are assessed in two U-world training sessions, separated by a 1-week interval [17,18]. With a standard test of 50 paw reaches in the U-world, the range of RPE scores is 0 to 50 right-paw reaches and the midpoint is 25. An individual mouse is considered to improve its RPE score in a second U-world test if it behaves the same as it did in the first U-world test or it is more biased in the same left-pawed or right-pawed direction as it was in its first U-world test. The improvement ratio (IR) is a binary measure and it is the fraction or proportion of the population that improves its RPE score in the second U-world training session. Since each inbred mouse strains is a single genotype, the IR of the inbred strain is equivalent to a measure of penetrance of paw-preference learning phenotype in 50 paw reaches by the population of individuals with that strains genotype. The weighted improvement ratio (WIR) is the average amount or degree of improvement in RPE score, averaged over all individuals that were tested whether they improved or not. WIR is a measure of how much a genotype does improve and it is similar to a measure of expressivity of the learning (improvement) phenotype in the individuals with that strains genotype.

The IR and WIR measures were determined from contemporary samples of the C57BL/6J, DBA/2J and CDS/Lay strains and compared to a sample of simulated nonlearner model mice with a zero or null training rate. Mice were assessed with 50 paw reaches in two training sessions in the U-world, separated by a 1-week interval. Distributions of RPE scores from the first U-world test with 50 reaches are shown in Figure 10 and the IR and WIR measures, derived from the second test, are summarized in Figure 11. The relationship between IR and WIR is intriguing because the linear trend line projects from the null model of the nonlearner mouse through the means of the three different biological mice and points at a perfect or constitutive behavior.

Simulation predicts the numerical values for the IR and WIR measures of paw preference for a nonlearner mouse that reaches but cannot learn a preference because it has a null training rate [18]. We expect that a null mutation or gene knock out that results in a null training rate would cause a mouse to reach randomly with its right and left paw, but the numerical value of IR and WIR would not be zero; the RPE scores will improve by chance in the retest of some individuals and their learning will be spurious. In contrast, a so-called perfect mouse is not a mouse that learns in the classic sense because the paw used for the first reach is used in all subsequent reaches and that individual would be described as having a constitutive paw preference. The only difference among mice with a constitutive paw preference might be whether some strains had both right-and left-pawed mice and other constitutive strains had only right- or left-pawed mice. We have not yet observed mice with a constitutive preference in wide surveys of different inbred strains as well as different species of mice [26].

We have previously used the relationship between penetrance and expressivity to assess threshold traits and right-left asymmetries of embryonic development [56]. One example is the asymmetrical right-left malformation response of mouse forelimbs to teratogenic insult. Change in penetrance and correlation with severity of expression of reduction deformities revealed a right to left gradient in the response to teratogens of mouse embryos; dramatic changes in slope of the correlation for the right and the left forelimbs between different mouse strains demonstrated that the right-left gradient was different between strains and, hence, it was an intrinsic property of the genotype rather than the forelimbs.

Simulations demonstrate that the patterns of paw preference in the U-world (Fig. 10) are predicted by strain differences in learning rate. Because the patterns are reliable and characteristic of the strains, we infer that differences in genes or interacting genes cause the strain-specific training rates and stochastic variation during reaching causes the predictable dynamic variation in the strain-specific

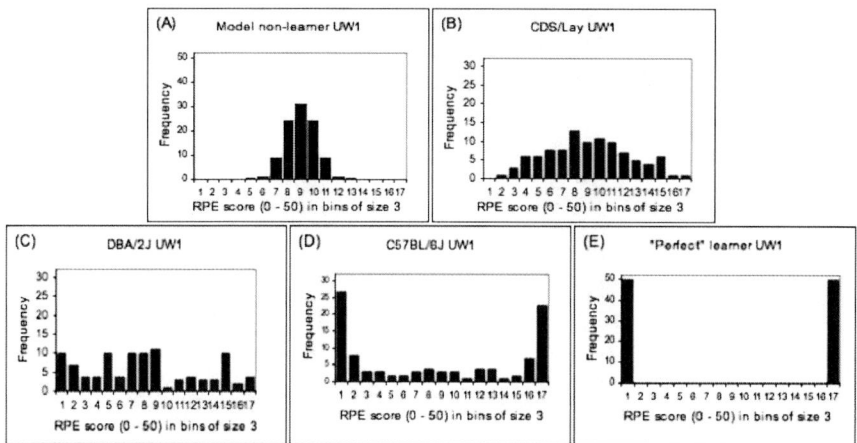

Fig. 10. Dynamic variation in patterns of hand preference in experimental mice compared with simulated model nonlearner and perfect learner mice. A contemporary sample of CDS/Lay (B), DBA/2J (C) and C57BL/6J (D) was allowed 50 reaches. Nonlearners (A) reach with right and left paw but have a null training rate of 0 memory units per successful reach; perfect learners (E) have a random first paw reach but use that hand in all subsequent reaches.

distributions of RPE scores. The respective population parameters of IR and WIR (Fig. 11) provide a measure of the binary proportion of individuals that improve in their retest and the average amount of improvement that is expressed by all the individuals that are retested. The relationship in Figure 11 among the biological mice is exactly predicted by the training rate. Therefore, the challenge we visualize for reductive genetics is to show how to exactly predict this training rate from the different deterministic genotypes of different mouse strains.

Figure 11 presents a more fundamental challenge to our understanding of the functional biology of paw preference behavior. How close can a biological mouse get to a nonlearner or a perfect learner? What is the biology that moves the behavior up and down on the line between null and perfect? Can the biology ever be found off the line? If the functional interactions of genes change the capacity of the mouse to learn the paw preference trait, are there small molecules (drugs, diet) that can also influence the trait? Opening the analysis to an assessment of penetrance and expressivity in a complex trait appears to open the analysis to an assessment of the emergent properties of the underlying elements. For the geneticist, the challenge is to determine what can change the distinctive learning rates, for example between the mutually exclusive genotypes of CDS/Lay, DBA/2J and C57BL/6J mice? The metaphor of biology lives on the line (Fig. 11) sets that task in motion and, at the same time, we are beginning to see how to construct the biology (as in Fig. 9) in order to go from a specified genotype to a predicted phenotype.

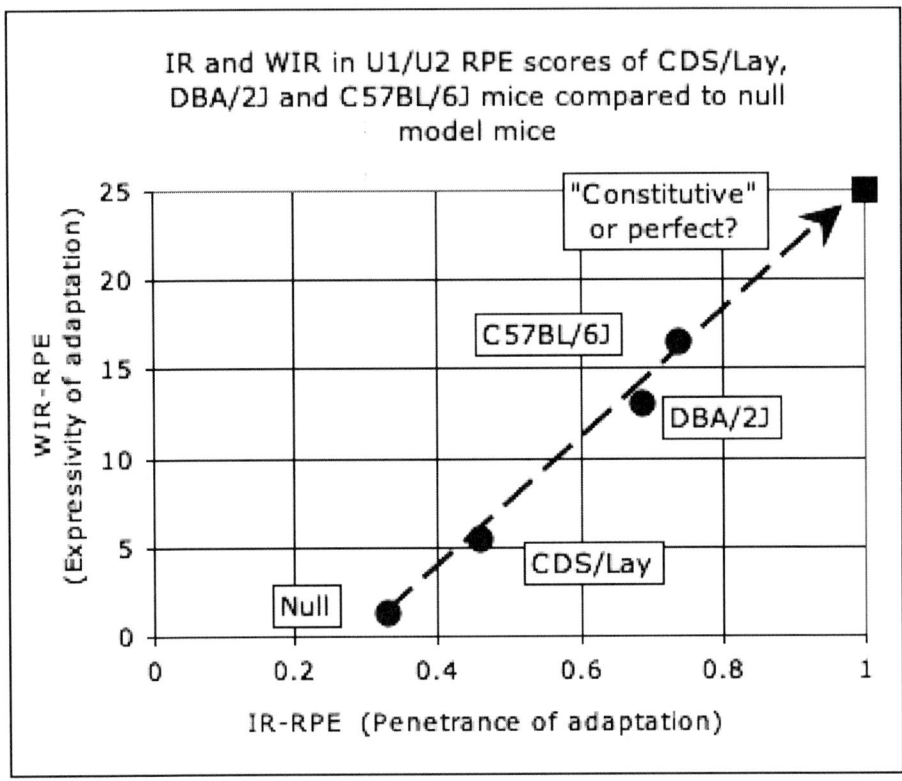

Fig. 11. Hand preference learning ability in mice is a genetically regulated adaptive behavior in response to reaching. A deterministic genotype-specific learning rate per successful hand reach predicts the diversity in the dynamic variation in RPE score from 50 hand reaches in the U-world; hence, it predicts the linear relationship between improvement ratio (IR) and weighted improvement ratio (WIR) from two U-world tests, separated by a one-week interval and allowing for memory consolidation. CDS/Lay, DBA/2J and C57BL/6J experimental mice are compared with numerically simulated nonlearner model mice and the linear trend line predicts the perfect learner model mice.

5 Conclusions

Noisy patterns of paw-preference scores of inbred mouse strains contain more information than what can be inferred from summaries of means and variances of the quantitative behavioral scores. Qualitative differences in the dynamic patterns of paw preference revealed the learning and memory process in the mouse behavior and reliability of the different patterns between different inbred mouse strains demonstrated the process is genetically regulated. Therefore, an adaptive behavior that depends on an individuals past experience is hiding behind the genes that regulate mouse paw preference. Simulations with a stochastic agent-based model weaned our analytical dependence off mean effects and focused our attention on the information in patterns of paw preference. The inherited deterministic trait is the learning rate per successful reach by which an individual mouse acquires its paw preference. Means and variances of samples of paw-preference scores are only approximations of the behavior and approximations obliterated the effect we are trying to study.

Individual mice receive information from the action of reaching and the resulting change in their paw preference reveals a sensitive, stochastic biological control system. Kinetic analysis uncovered four elements of the system involving time: the time for consolidation of memory of previous training reaches, decay of memory after training, increase in amount of memory with increase in number of training reaches, and different amounts of memory achieved by different mouse strains with those training reaches. Gene activity is essential for this memory process because a protein synthesis inhibitor can block consolidation of memory after training.

Stochastic agent-based modeling captured biologically realistic features in paw-reaching behavior. Model mice retain information from their training and it has a consequence on subsequent reaching events. Simulations with the model showed how individual-to-individual biological variation arises between genetically identical mice that have the same training rate. Variation in preference scores among genetically identical individuals emerges from the stochastic nature of paw reaching. A probabilistic change in paw reaching occurs from reach to reach in an individual mouse in a training session due to the interaction between the individual and its environment as well as between consecutive training sessions. Simulations with model mice matched the kinetics of paw-preference learning and predicted the limits of learning for different mouse strains; at the same time, simulations tell us there is no unaccounted factor in the model because the parameters of learning are the same in a relearning situation. Since paw usage is a genetically regulated probabilistic choice rather than a deterministic preference, a mouse is able to adapt to any world bias and still maintain an ability to detect and respond to further changes in world bias, regardless of the amount of previous training. Some mouse strains simply adapt better than others.

Improvement ratio (IR) and weighted improvement ratio (WIR) from two consecutive U-world tests are new measures of paw preference behavior. They are the respective fraction of individuals that improve their preference and the amount of learning of the tested population. In concept, they represent the penetrance and expressivity of the amount of learning by different genotypes of mice. The apparent linear relationship between IR and WIR (Fig. 11) shows that the biology of this adaptive behavior lives on the line between null and perfect, that is between a random paw reaching with no apparent learning and a behavior that is an inflexible constitutive paw preference. We can ask whether no learning and constitutive preference are biologically possible and whether any genotype of mouse is off the line. The IR and WIR relationship may be a method to identify interactions between different genes and environmental treatments that qualitative alter the system of leaning and memory in this behavior.

Genetic diversity in learning rate focuses attention in the direction of the biology and evolutionary significance of the adaptive behavior and away from the curiosity of a right-paw and a left-paw preference. Our biological question is whether learning rate in this behavior is the result of single genes in a developmental pathway or the emergent systems property of dynamic networks of interacting genes. If there are interacting genes, what are they and what are their functions? If genetic diversity is maintained in natural populations of mice, the evolutionary question is whether training rate provides an individual with a selective advantage or simply an unintended consequence of other regulatory genes and functions? Therefore, learning and memory is a practical framework to study the systems structure and genetic architecture of the mouse behavior. Our study of the noisy behavior patterns has reinforced the statement that 'the organism is determined neither by its genes nor by its environment nor even by the interaction between them, but bears a significant mark of random processes' [57].

Finally, the genetically regulated mouse model provides a model system to interpret human hand preference behavior and to assess some of its neglected properties, such as "improvement" in degree of lateralization with age in children. The mouse model strongly supports an interpretation that hand preference may not be due to brain lateralization, but instead it reflects a positive feedback mechanism in which hand usage simply makes that hand more preferred in future actions [58]. We expect that deeper analysis of the mouse model will guide the analysis of the human behavior and a critical evaluation of its mythology.

Acknowledgments

We thank the editors for their invitation to contribute our work and for their patience. Behavior geneticists usually ignore the adage [59]: 'Genes dont cause behaviors. Sometimes they influence them.' Collaborative interaction with Dr. Andre S. Ribeiro and his team in the Computational Systems Biology Research Group, Tampere University of Technology, turned our attention to information processing in the adaptive behavior of mouse paw preference and revealed new insight into the genetics and evolutionary significance of the phenotypic diversity. We are beginning to see how genes can sometimes influence hand preference behavior.

References

1. Jacob, F., Monod, J.: On the regulation of gene activity. Cold Spring Harbor Symp. Quant. Biol. 26, 193–211 (1961)
2. Bishop, D.V.M.: Handedness and Developmental Disorder. MacKieth Press, London (1990)
3. Coren, S.: The Left-hander Syndrome. In: The Causes and Consequences of Left-handedness. The Free Press, Maxwell Macmillan International, New York (1992)

4. Manus, C.: Right Hand, Left Hand. In: Manus, C. (ed.) The Origins and Asymmetry in Brains, Bodies, Atoms and Culture. Harvard University Press, Cambridge (2002)
5. Tommasi, L.: Mechanisms and functions of brain and behavioural asymmetries. Phil. Trans. R. Soc. B 364, 855–859 (2009)
6. Rodriguez, A., Kaakinen, M., Moilanen, I., Taanila, A., McGough, J.J., Loo, S., Jarvelin, M.-R.: Mixed Handedness is linked to mental health problems in children and adolescents. Pediatrics 125, e340–e348 (2010)
7. Medland, S.E., Duffy, D.L., Wright, M.J., Geffen, G.M., Hay, D.A., Levy, F., van-Beijsterveldt, C.E.M., Willemsen, G., Townsend, G.C., White, V., Hewitt, A.W., Mackey, D.A., Bailey, J.M., Slutske, W.S., Nyholt, D.R., Treloar, S.A., Martin, N.G., Boomsma, D.I.: Genetic influences on handedness: Data from 25,732 Australian and Dutch twin families. Neuropsychologia 47, 330–337 (2009)
8. Llaurens, V., Raymond, M., Faurie, C.: Why are some people left-handed? An evolutionary perspective. Phil. Trans. R. Soc. B 364, 881–894 (2009)
9. Schaafsma, S.M., Riedstra, B.J., Pfannkuche, K.A., Bouma, A., Groothuis, T.G.G.: Epigenesis of behavioural lateralization in humans and other animals. Phil. Trans. R. Soc. B 364, 915–927 (2009)
10. Corballis, M.C.: Review: the evolution and genetics of cerebral asymmetry. Phil. Trans. Roy. Soc. B 364, 867–879 (2009)
11. Crawley, J.N.: What's Wrong With My Mouse? Behavioral Phenotyping of Transgenic and Knockout Mice. Wiley-Liss, New York (2000)
12. Beck, J.A., Lloyd, S., Hafesparast, M., Lennon-Pierce, M., Eppig, J.T., Festing, M.F.W., Fisher, E.M.C.: Genealogies of mouse inbred strains. Nature Genet. 24, 23–25 (2000)
13. Grubb, S.C., Maddatu, T.P., Bult, C.J., Bogue, M.A.: Mouse phenome database. Nucleic Acids Res. 37, 720–730 (2009)
14. Wahlsten, D., Rustay, N.R., Metten, P., Crabbe, J.C.: In search of a better mouse test. Trends Neurosci. 26, 132–136 (2003)
15. Biddle, F.G., Eales, B.A.: Mouse genetic model for left-right hand usage: Context, direction, norms of reaction, and memory. Genome 42, 1150–1166 (1999)
16. Biddle, F.G., Eales, B.A.: Hand-preference training in the mouse reveals key elements of its learning and memory process and resolves the phenotypic complexity in the behaviour. Genome 49, 666–677 (2006)
17. Ribeiro, A.S., Lloyd-Price, J., Eales, B.A., Biddle, F.G.: Dynamic agent-based model of hand-preference behavior patterns in the mouse. Adapt. Behav. 18, 116–131 (2010)
18. Ribeiro, A.S., Eales, B.A., Biddle, F.G.: Learning of paw preference in mice is strain dependent, gradual and based on short-term memory of previous reaches. Animal Behaviour (2010), doi:10.1016/j.anbehav.2010.10.014
19. Collins, R.L.: On the inheritance of handedness. I. Laterality in inbred mice. J. Hered. 59, 9–12 (1968)
20. Collins, R.L.: On the inheritance of handedness. II. Selection for sinistrality in mice. J. Hered 60, 117–119 (1969)
21. Weiss, K.M., Buchanan, A.V.: Evolution by phenotype, a biomedical perspective. Perspect. Biol. Med. 46, 159–182 (2003)
22. Betancur, C., Neveu, P.J., Le Moal, M.: Strain and sex differences in the degree of paw preference in mice. Behav. Brain Res. 45, 97–101 (1991)
23. Signore, P., Chaoui, M., Nosten-Bertand, M., Perez-Diaz, F., Marchaland, C.: Handedness in mice: comparison across eleven inbred strains. Behav. Genet. 53, 421–429 (1991)

24. Biddle, F.G., Coffaro, C.M., Ziehr, J.E., Eales, B.A.: Genetic variation in paw preference (handedness) in the mouse. Genome 36, 935–943 (1993)
25. Takeda, S., Endo, A.: Paw preference in mice; a reappraisal. Physiol. Behav. 53, 727–730 (1993)
26. Biddle, F.G., Eales, B.A.: The degree of lateralization of paw usage (handedness) in the mouse is defined by three major phenotypes. Behav. Genet. 26, 391–406 (1996)
27. Collins, R.L.: On the inheritance of direction and degree of asymmetry. In: Glick, S.D. (ed.) Cerebral Lateralization in Nonhuman Species, pp. 41–71. Academic Press, Orlando (1985)
28. Collins, R.L.: Reimpressed selective breeding for lateralization of handedness in mice. Brain Res 564, 194–202 (1991)
29. Wahlsten, D.: Evaluating genetic models of cognitive evolution and behaviour. Behav. Process. 35, 183–194 (1996)
30. Ward, R., Tremblay, L., Lassonde, M.: The relationship between callosal variation and lateralization in mice is genotype-dependent. Brain Res. 424, 84–88 (1987)
31. Bulman-Fleming, B., Wainwright, P.E., Collins, R.L.: The effects of early experience on callosal development and functional lateralization in pigmented BALB/c mice. Behav. Brain Res. 50, 31–42 (1992)
32. Gruber, D., Waanders, R., Collins, R.L., Wolfer, D.P., Lipp, H.-P.: Weak or missing paw lateralization in a mouse strain (I/Ln) with congenital absence of the corpus callosum. Behav. Brain Res. 46, 9–16 (1991)
33. Lipp, H.-P., Collins, R.L., Hausheer-Zarmakupi, Z., Leisinger-Trigona, M.-C., Crusio, W.E., Nosten-Bertrand, M., Signore, P., Schwegler, H., Wolfer, D.P.: Paw preference and intra-/infrapyramidal mossy fibers in the hippocampus of the mouse. Behav. Genet. 26, 379–390 (1996)
34. Neveu, P.J., Barneoud, P., Vitiello, S., Betancur, C., Le Moal, M.: Brain modulation of the immune system: association between lymphocyte responsiveness and paw preference in mice. Brain Res. 457, 392–394 (1988)
35. Neveu, P.J., Betancur, C., Vitiello, S., Le Moal, M.: Sex-dependent association between immune function and paw preference in two substrains of C3H mice. Brain Res. 559, 347–351 (1991)
36. Denenberg, V.H., Sherman, G.F., Schrott, L.M., Rosen, G.D., Galaburda, A.M.: Spatial learning, discrimination learning, paw preference and neocortical ectopias in two autoimmune strains of mice. Brain Res. 562, 98–114 (1991)
37. Denenberg, V.H., Mobraaten, L.E., Sherman, F.G., Morrison, L., Schrott, L.M., Waters, N.S., Rosen, G.D., Behan, P.O., Galaburda, A.M.: Effects of the autoimmune uterine/maternal environment upon cortical ectopias, behavior and autoimmunity. Brain Res. 563, 114–122 (1991)
38. Roubertoux, P.L., Le Roy, I., Tordjman, S., Cherfou, A., Migliore-Samour, D.: Analysis of quantitative trait loci for behavioral laterality in mice. Genetics 163, 1023–1030 (2003)
39. Collins, R.L.: When left-handed mice live in right-handed worlds. Science 187, 181–184 (1975)
40. Biddle, F.G., Eales, B.A.: Lateral asymmetry of paw usage: phenotypic survey of constitutive and experience-conditioned behaviours among common strains of the mouse. Genome 44, 539–548 (2001)
41. Biddle, F.G., Jones, D.A., Eales, B.A.: A two-locus model for experience-conditioned direction of paw usage in the mouse is suggested by dominant and recessive constitutive paw usage behaviours. Genome 44, 872–882 (2001)

42. Kaern, M., Elston, T.C., Blake, W.J., Collins, J.J.: Stochasticity in gene expression: from theories to phenotypes. Nature Rev. Genet. 6, 451–464 (2005)
43. Yu, J., Xiao, J., Ren, X., Lao, K., Xie, X.S.: Probing gene expression in live cells, one protein molecule at a time. Science 311, 1600–1603 (2006)
44. Samoilov, M.S., Price, G., Arkin, A.P.: From fluctuations to phenotypes: the physiology of noise. Science STKE 366, re17 (2006)
45. Suel, G., Garcia-Ojalvo, J., Liberman, L., Elowitz, M.B.: An excitable gene regulatory circuit induces transient cellular differentiation. Nature 440, 545–550 (2006)
46. Arkin, A.P., Ross, J., McAdams, H.H.: Stochastic kinetic analysis of developmental pathway bifurcation in phage-infected Escherichia coli cells. Genetics 149, 1633–1648 (1998)
47. Wahlsten, D.: Experimental design and statistical inference. In: Crusio, W.E., Gerlai, R.T. (eds.) Handbook of Molecular-Genetic Techniques for Brain and Behavior Research. Techniques in the Behavioral and Neural Sciences, vol. 13, pp. 41–57. Elsevier Science, B.V., Amsterdam (1999)
48. Nakagawa, S., Cuthill, I.C.: Effect size, confidence interval and statistical significance: a practical guide for biologists. Biol. Rev. 82, 591–605 (2007)
49. Scriver, C.R., Waters, P.J.: Monogenic traits are not simple: lessons from phenylketonuria. Trends Genet. 15, 267–272 (1999)
50. Scriver, C.R.: The PAH gene, phenylketonuria, and a paradigm shift. Human Mutation 28, 831–845 (2007)
51. Morgan, T.M., Krumholz, H.M., Lifton, R.P., Spertus, J.A.: Nonvalidation of reported genetic risk factors for acute coronary syndrome in a large-scale replication study. J. Am. Med. Assoc. 297, 1551–1561 (2007)
52. Weiss, K.M.: Tilting at quixotic trait loci (QTL): An evolutionary perspective on genetic causation. Genetics 179, 1741–1756 (2008)
53. Anderson, P.W.: More is different. Science 177, 393–396 (1972)
54. Francks, D., DeLisi, L.E., Fisher, S.E., Laval, S.H., Rue, J.E., Stein, J.F., Monaco, A.P.: Confirmatory evidence for linkage of relative hand skill to 2p12-q11. Am. J. Hum. Genet. 72, 499–502 (2003)
55. Medland, S.E., Duffy, D.L., Spurdle, A.B., Wright, M.J., Geffen, G.M., Montgomery, G.W., Martin, N.G.: Opposite effects of androgen receptor CAG repeat length on increased risk of left-handedness in males and females. Behav. Genet. 35, 735–744 (2005)
56. Biddle, F.G., Mulholland, L.R., Eales, B.A.: Penetrance and expressivity of acetazolamide-ectrodactyly provide a method to define a right-left teratogenic gradient that differs between the C57BL/6J and WB/ReJ mouse strains. Teratology 47, 603–612 (1993)
57. Lewontin, R.: The Triple Helix: Gene, Organism, and Environment. Harvard University Press, Cambridge (2000)
58. McManus, I.C., Sik, G., Cole, D.R., Mellon, A.F., Wong, J., Kloss, J.: The development of handedness in children. Brit. J. Dev. Psych. 6, 257–273 (1988)
59. Sapolsky, R.M.: Monkeyluv: And Other Essays on Our Lives as Animals. Scribner, New York (2005)

Stochastic Gene Expression and the Processing and Propagation of Noisy Signals in Genetic Networks

Daniel A. Charlebois[1,2], Theodore J. Perkins[3,4], and Mads Kaern[1,2,5]

[1] Ottawa Institute of Systems Biology, University of Ottawa,
451 Smyth Road, Ottawa, Ontario, K1H 8M5, Canada
[2] Department of Physics, University of Ottawa,
150 Louis Pasteur, Ottawa, Ontario, K1N 6N5, Canada
[3] Ottawa Hospital Research Institute,
501 Smyth Road, Ottawa, Ontario, K1H 8L6, Canada
[4] Department of Biochemistry, Microbiology and Immunology, University of Ottawa,
451 Smyth Road, Ottawa, Ontario, K1H 8M5, Canada
[5] Department of Cellular and Molecular Medicine, University of Ottawa,
451 Smyth Road, Ottawa, Ontario, K1H 8M5, Canada

Abstract. Over the past few years, it has been increasingly recognized that stochastic mechanisms play a key role in the dynamics of biological systems. Genetic networks are one example where molecular-level fluctuations are of particular importance. Here stochasticity in the expression of gene products can result in genetically identical cells displaying significant variation in biochemical or physical attributes. This variation can influence individual and population-level fitness.

Cells also receive noisy signals from their environments, perform detection and transduction with stochastic biochemistry. Several mechanisms, including cascades and feedback loops, allow the cell to manipulate noisy signals and maintain signal fidelity. Furthermore through a biochemical implementation of Bayes's rule, it has been shown that genetic networks can act as inference modules, inferring from intracellular conditions the likely state of the extracellular environment.

Keywords: stochastic gene expression, fitness, genetic networks, signal processing and propagation, Bayesian inference.

1 Introduction

Genetic networks, defined as ensembles of molecules and interactions that control gene expression, produce and regulate cellular dynamics. At a fundamental level, a gene is information encoded in a sequence of nucleotides. This information is processed by the machinery of the cell to execute the instructions it contains. Understanding the process by which this information is produced, processed, and propagated is vital for understanding cellular behaviour.

Advancements in experimental techniques for empirically measuring gene expression in single cells, as well as in corresponding theoretical methods, have enabled the rigorous design and interpretation of experiments that provide incontrovertible proof that there are important endogenous sources of stochasticity (randomness) that drive biological processes [51]. For example, heterogeneity within a population of a single cell type can be measured experimentally using flow-cytometry analysis, a technique commonly employed for counting and examining the chemical and physical properties of cells. Specifically, one can obtain within a few seconds a histogram of a given protein in individual cells across a large cell population. Within the histogram, the abundance of the protein in the cells with the lowest and highest expression level typically differs by orders of magnitude; this spread far exceeds signal measurement noise [15].

The stochastic expression of gene products (mRNA and protein) is important for human health and disease. Take for example the development of drug resistance during chemotherapy. When the drug Imatinib is used to treat chronic myeloid leukemia, the disease recurs with a frequency of 20-30 % [14]. Even though numerous genetic mutations have been shown to render the drug ineffective [27,23,63], in two-thirds of cases no mutations have been found [14]. Instead, elevated levels of survival pathway proteins in Imatinib-resistant leukaemia cell lines were detected [37]. The rapid rate of resistance development, its dose dependence and high frequency of upregulation of the correct pathways are consistent with non-genetic heterogeneity, that is, variation in gene expression across a population of genetically identical cells. This mechanism generates enduring outlier cells with distinct phenotypes (i.e. any observable biochemical or physical attributes), some of which may be subject to selection.

Cells sense and process information using biochemical networks of interacting genes and proteins [29]. At a specific point of the network (input) a signal is detected and then is propagated to modulate the activity or abundance of other network components (output). In order to process information reliably, the cell requires a high degree of sensitivity to the input signal but a low sensitivity to random fluctuations in the transmitted signal. However, the signals that a cell receives from its environment and propagates through its genetic network are noisy [29,43]. Understanding how this noise is processed and propagated in gene networks is crucial for understanding signal fidelity in natural networks and designing noise-tolerant gene circuits [42]. For example, several network motifs allow for amplification or attenuation of noisy signals [4,8,18,28,38,42,44,49,52,57]. Additionally, it has been shown that genetic networks can be used by cells to infer the likely state of their stochastic external environment from noisy intercellular conditions [32,43].

The chapter is organized as follows: Section 2 presents the process of gene expression, the inherent stochasticity in this process, and common measures of noise. Some background for the deterministic and stochastic modelling and simulation of gene expression, as well as a comparison between these two methods, is provided in Section 3. Section 4 introduces mechanisms, namely genetic cascades and feedback loops, that enable the cell to process and propagate noisy intracellular and extracellular signals. In Section 5, the relationship between noise and fitness is explored. Specifically, the stochastic expression of stress-related genes and bet-hedging cell populations are discussed, and corresponding models and simulations are presented. The final section (Section 6), illustrates how genetic networks can infer the likely state of their extracellular environment through a biochemical implementation of Bayes' rule.

2 Gene Expression and Stochasticity

A gene is a specific sequence of nucleotides encoded in the DNA. Gene expression is the process by which a gene is transcribed and translated to produce messenger RNA (mRNA) and protein, respectively. To initiate transcription, an RNA polymerase (RNAp) must recognize and bind to the promoter region of the gene. Promoters have regulatory sites to which transcription factors can bind to either activate or repress gene transcription. The promoter is followed by the coding sequence, which is transcribed by the RNAp into an mRNA molecule. Transcription stops when the RNAp reaches a termination sequence and unbinds from the DNA. Next, translation ensues wherein ribosomes read the mRNA sequence, and for each codon, a corresponding amino acid is added to a polypeptide chain (a.k.a. a protein). After post-translational processing, the protein becomes capable of performing its specific tasks.

A model of the process of expressing a single gene is shown in Figure 1. Although this depiction is simple compared to the true complexity of gene expression, it captures the essential features including the synthesis of mRNA (M) from a single gene promoter (A) (at a rate s_A), the synthesis of protein (P) from mRNA templates (rate s_P), and the decay of mRNA and protein molecules (rates δ_M and δ_P respectively). Although more complex models of gene expression have been developed (e.g. [34,47,50,56]), the simple model depicted in Figure 1 is sufficient for the purpose of this chapter.

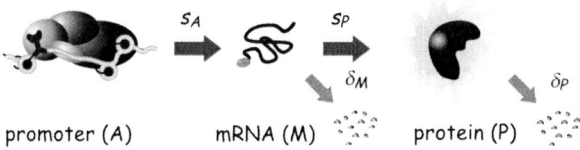

Fig. 1. A simple model for the expression of a single gene (each step represents several biochemical reactions). All steps are modelled as first-order reactions with the indicated rate constants (units of inverse time) associated with these steps.

The expression of gene products is a noisy process [51,30,31,35,41]. The term 'noise' when used in the context of gene expression is a broad reference to the observed variation in protein content among apparently identical cells exposed to the same environment [21]. This noise can be divided up into extrinsic and intrinsic components. Extrinsic noise can be generally defined as fluctuations and variability that arise in a system due to disturbances originating from its environment, and therefore depends on how the system of interest is defined [53]. Extrinsic gene expression noise arises from several sources including: the metabolic state of the cell, cell-cycle phase, cell age, variability in upstream signal transduction, and the external cellular environment [19,21,30,32,35,42,45,46,56,62]. Intrinsic expression noise refers to the multistep processes that lead to the synthesis and degradation of mRNA and protein molecules which are inherently stochastic due to the underlying binding events which occur as a result of the random collisions between small numbers of molecules (e.g. the binding of transcription factors to one or two copies of a gene) [30].

Several noise measures are used to quantify the degree of heterogeneity in gene expression. The most common is the relative deviation from the average, which is determined by the ratio of the standard deviation σ to the mean μ. In this chapter, noise η refers to this ratio. Another measure of noise, known as the 'fano factor' ($\phi = \sigma^2/\mu$), can be used to uncover trends that might otherwise be obscured by the characteristic $1/\sqrt{\mu}$ scaling of the noise [30,57].

3 Modelling Gene Expression

Biological systems can be modelled at multiple scales, from detailed physical descriptions of molecular interactions to phenomenological representations of populations of organisms. Here we present the approximate ordinary differential equation (ODE) approach and the exact stochastic method to simulate the phenomenological model of gene expression shown in Figure 1.

3.1 Deterministic Modelling

Traditionally, the time evolution of a chemical system is modelled as a deterministic process using a set of ODEs. This approach is based on the empirical law of mass action, which provides a relation between reaction rates and molecular concentrations [60]. Generally, the instantaneous rate of a reaction is directly proportional to the concentration (which is in turn proportional to mass). In the deterministic description of the model shown in Figure 1, the cellular mRNA and protein concentrations ($[M]$ and $[P]$, respectively) are governed by the macroscopic rate equations

$$\frac{d[M]}{dt} = s_A - \delta_M[M], \tag{1}$$

$$\frac{d[P]}{dt} = s_P[M] - \delta_P[P], \tag{2}$$

where the terms $\delta_M[M]$ and $\delta_P[P]$ are the degradation rates for mRNA and protein, respectively; the term $s_p[M]$ is the rate of protein synthesis, and mRNA production occurs at a constant rate (s_A) due to the presence of a single promoter. The steady-state concentrations are given by

$$[M^s] = \frac{s_A}{\delta_M}, \tag{3}$$

$$[P^s] = \frac{[M]s_P}{\delta_P} = \frac{s_A s_P}{\delta_M \delta_P}, \tag{4}$$

and are related to the average steady-state number of M and P (M^s and P^s, respectively) by the cell volume V.

Note that the deterministic mathematical model (Eqs. (1) and (2)) was obtained by treating each step as a first-order chemical reaction and applying the law of mass action. The law of mass action was developed to describe chemical reactions under conditions where the number of each chemical species is so large that concentrations can be approximated as continuous variables without introducing significant error [53].

In order for the deterministic approach to provide a valid approximation of the exact stochastic description, the system size must be large in terms of the numbers of each species and the system volume (e.g., here large s_A and V so that the number of expressed mRNA and protein molecules is high with the ratio s_A/V remaining constant) [30]. When this condition is not satisfied, the effects of molecular noise can be significant. The high molecular number condition is not satisfied for gene expression, due to low copy number of genes, mRNAs, and transcription factors within the cell [64].

When the deterministic ODEs presented in Eqs. (1) and (2) are numerically simulated (e.g. via a variable step Runge-Kutta method), the resulting trajectory can in certain parameter regimes capture the mean behavior of the cells. They cannot, however, capture the fluctuations about the mean and therefore the resulting probability distributions (Fig. 2). Futhermore, when reaction rates

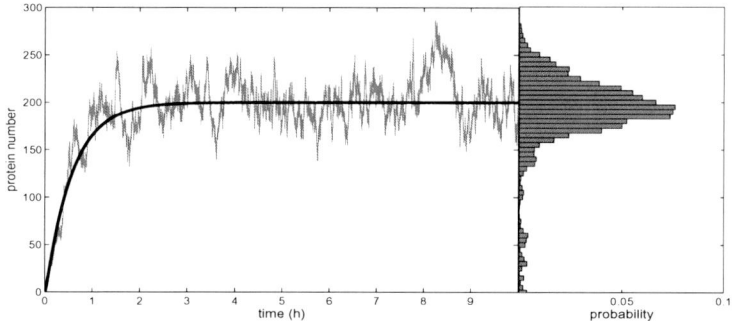

Fig. 2. Time series of protein number generated by deterministic and stochastic simulations (black and gray curves, respectively). The histogram in the right-hand panel corresponds to the stochastic simulation and shows the probability that a cell will have a given intracellular protein level. Parameters were set to (units s^{-1}): $s_A = 0.02$, $s_P = 0.05$, $\delta_M = 0.0005$, and $\delta_P = 0.01$.

depend nonlinearly on randomly fluctuating components, macroscopic rate equations may be far off the mark even in their estimates of averages [40].

3.2 Stochastic Modelling

Due to the importance of noise in many biological systems, models involving stochastic formulations of chemical kinetics are increasingly being used to simulate and analyze cellular control systems [26]. In many cases, obtaining analytical solutions for these models is not feasible due to the intractability of the corresponding system of nonlinear equations. Thus, Monte Carlo (MC) simulation procedures for the number of each molecular species are commonly employed. Among these procedures, the Gillespie stochastic simulation algorithm (SSA) is the *de-facto* standard for simulating biochemical systems in situations where a deterministic formulation may be inadequate [24,25].

In the *direct method* Gillespie SSA, M chemical reactions $\{R_1, \ldots, R_M\}$ characterised by numerical reaction parameters $c_1, \ldots, c_M$ among N chemical species $X_1, \ldots, X_N$, are simulated one reaction event at a time. The fundamental hypothesis of the stochastic formulation of chemical kinetics is that the average probability of a given reaction i, occurring in the next infinitesimal time interval, dt, is given by $a_i dt$. Here, a_i is the reaction propensity obtained by multiplying c_i by the number of reactants (for first order reactions) or reactant combinations (for second order and higher reactions) h_i available for reaction R_i. The next reaction to occur (index μ) and its timing τ are determined by calculating the M reaction propensities $a_1, \ldots, a_M$ to obtain an appropriately weighted probability for each reaction. The SSA determines when ($\tau = \ln(1/r_1)/a_0$) and which ($\min\{\mu \mid \sum_{i=1}^{\mu} a_i \geq r_2 a_0\}$) reaction will occur next, using uniformly distributed random numbers r_1 and r_2, and the sum of the reaction propensities a_0.

The direct method Gillespie SSA can be implemented via the following pseudocode [24,25]:

1: **if** $t < t_{end}$ and $a_0 = \sum_{i=1}^{M} a_i \neq 0$ **then**
2:   **for** $i = 1, M$ **do**
3:     Calculate a_i and $a_0 = \sum_{v=1}^{i} a_v$
4:   **end for**
5:   Generate r_1 and r_2
6:   Determine τ and μ
7:   Set $t = t + \tau$
8:   Update $\{X_i\}$
9: **end if**

The following reaction equations are required to stochastically simulate the model of gene expression under consideration (Fig. 1)

$$A \xrightarrow{s_A} A + M \tag{5}$$

$$M \xrightarrow{s_P} M + P \tag{6}$$

$$M \xrightarrow{\delta_M} \oslash \qquad (7)$$
$$P \xrightarrow{\delta_P} \oslash \qquad (8)$$

Eqs. (5) and (6) respectively describe the transcription and translation processes. The degradation of M and P are accounted for by Eqs. 7 and 8, respectively.

The advantage of using a stochastic framework to simulate the present model of gene expression can be seen in Figure 2. Specifically, the stochastic method captures not only the mean protein concentration, but also the fluctuations in protein abundance. These fluctuations provide the information necessary for the histograms that describe the probability that a cell will have a given level of a particular molecular species, and can play a significant role in cellular dynamics.

4 Processing and Propagation of Noisy Signals

The genetic program within a living cell is encoded by a complex web of biochemical interactions between gene products. The proper execution of this program depends on the propagation of signals from one gene to the next. This process may be hindered by stochastic fluctuations arising from gene expression. Furthermore it has been found that gene expression noise not only arises from intrinsic fluctuations, but also from noise transmitted from the expression of upstream genes [42]. We now consider how noise can be processed and propagated in genetic networks.

4.1 Cascades

A common regulatory motif, especially in development, is a transcriptional cascade where each gene (A_i) influences the expression of a subsequent gene (A_{i+1}) to form a cascade (Fig. 3 Inset) [44]. Experimental studies have shown that variability can be transmitted from an upstream gene to a downstream gene, adding substantially to the noise inherent in the downstream gene's expression [42,49].

Using a reduced version of the model of gene expression presented in Figure 1, where transcription and translation are combined into a single step, we model a generic linearised genetic cascade as follows. The input signal for the cascade is provided by A_0, which itself is constitutively expressed to produce a protein P_0 and described by the following reactions

$$A_0 \xrightarrow{s_{P_0}} A_0 + P_0 \qquad (9)$$
$$P_0 \xrightarrow{\delta_{P_0}} \oslash \qquad (10)$$

The protein expression dynamics P_i of the subsequent genes A_i (where i $\in \{1,\ldots,N\}$, and N is the total number of genes) in the cascade are modelled as follows

$$A_i + P_{i-1} \xrightarrow{s_{P_i}} A_i + P_i + P_{i-1} \qquad (11)$$
$$P_i \xrightarrow{\delta_{P_i}} \oslash \qquad (12)$$

The protein expression of the genes $A_1 - A_N$ in the cascade are each subject to the stochastic fluctuations in the previous gene's expression. Therefore the noise in protein number, for the same mean expression, increases with each subsequent step in the cascade (Fig. 3 - includes parameters).

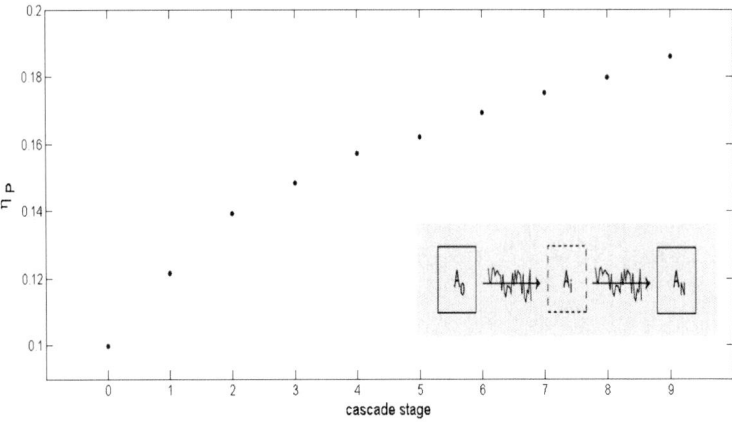

Fig. 3. Propagation of noise in a genetic cascade. The noise in protein number (η_P) is plotted against the cascade stage. Parameters were set as follows (units s^{-1}): $k_{P_0} = k_{P_i} = 100$ and $d_{P_0} = d_{P_i} = 1$ and the simulation was run for 100000 s in order to obtain accurate statistics. Inset shows a schematic of a generic linearised stochastic cascade where each gene (A_i) influences the expression of the subsequent gene in the cascade.

Genetic cascades can produce a wide range of dynamics in addition to those presented in this section. For example, it has been shown that genetic cascades can be either 'fluctuation-unbounded' (as in Fig. 3) or 'fluctuation-bounded' (i.e. expression noise moves towards some asymptotic limit as the size of the cascade is increased) [59]. Furthermore, longer genetic cascades can actually function to filter out rapid fluctuations at the expense of amplifying noise in the timing of propagated signals [59]. To perform this function, the cascade must not only be fluctuation-bounded, but must also be intrinsically less noisy than the input signal.

4.2 Feedback Loops

Feedback loops, in which a protein regulates its own transcription, play an important regulatory role in many genetic networks [38,44]. Positive feedback loops (e.g. where a protein activates its own expression) can act as noise amplifiers [38], whereas negative feedback loops (e.g. where a protein represses its own expression) can act to suppress noise [8,18,54]. Specifically, negative feedback can reduce the effects of noise because fluctuations above and below the mean are pushed back towards the mean [4,8,18,52,57]. Here we provide a simple example of relative noise amplification and attenuation in genetic feedback loops.

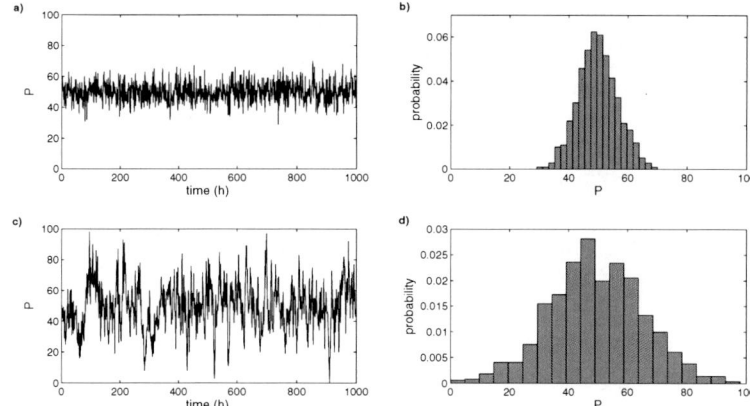

Fig. 4. Stochastic simulations of negative and positive feedback networks. Protein (P) time series and corresponding probability histograms of negative (a,b) and positive (c,d) auto-regulatory systems (Eqs. (13)-(16) and Eqs. (17)-(21), respectively). Note the increase in variability about the same mean when positive auto-regulation is compared to negative auto-regulation. Parameters are given in the text.

Again using the reduced version of the model of gene expression presented in Section 4.1, but where the protein P represses its own formation, we obtain a simple example of a network with negative auto-regulation [38]. The reactions are as follows

$$A + P \xrightarrow{k_1} AP \qquad (13)$$

$$AP \xrightarrow{k_2} A + P \qquad (14)$$

$$A \xrightarrow{s_P} A + P \qquad (15)$$

$$P \xrightarrow{\delta_P} \varnothing \qquad (16)$$

Here, Eqs. (13) and (14) respectively describe the binding and unbinding of P with a promoter A, Eq. (15) the production of P which occurs only when the promoter is not bound to P, and Eq. (16) the degradation of P. The reaction parameters were set as follows: $k_1 = 4\ mol^{-1}h^{-1}$, $k_2 = 100\ h^{-1}$, $s_P = 150\ mol^{-1}h^{-1}$, and $\delta_P = 1\ h^{-1}$. The protein time series and corresponding probability histogram are shown respectively in Figure 4a and 4b.

The corresponding positive auto-regulation system, where protein production occurs at a higher rate (than basal) when P is bound to A, can be described by the following reactions

$$A + P \xrightarrow{k_1} AP \qquad (17)$$

$$AP \xrightarrow{k_2} A + P \qquad (18)$$

$$A \xrightarrow{b_P} A + P \qquad (19)$$

$$AP \xrightarrow{s_P} AP + P \tag{20}$$

$$P \xrightarrow{\delta_P} \oslash \tag{21}$$

These equations are similar to those describing negative auto-regulation (Eqs. (13)-(16)) except that Eq. (19) describes basal protein production (which is required for activation) and Eq. (20) the promoter bound production of P. Here the parameters were set to: $k_1 = 1\ mol^{-1}h^{-1}$, $k_2 = 100\ h^{-1}$, $b_P = 3\ mol^{-1}h^{-1}$ $s_P = 147\ mol^{-1}h^{-1}$, and $\delta_P = 1\ h^{-1}$. Note the increase in noise in the protein time series and histogram (Fig. 4c and 4d, respectively) relative to the negative feedback case (Fig. 4a and 4b).

It is important to note that many dynamics not discussed in the present section can result from the manner in which a genetic network propagates and processes signals. For example, in the presence of noise, positive feedbacks can behave as a switch, eventually flipping the gene from an 'off' to an 'on' state [20,44]. Furthermore, negative feedback loops can control speed of response to intra or extra-cellular events [48] and lead to oscillations in the expression of a gene product [7]. Feedback loops have also been shown capable of shifting the frequency of gene expression noise such that the effect on noise behaviour of downstream gene circuits within a cascade may be negligible, thus acting as noise filters [54].

5 Noise and Fitness

Heterogeneity in a cell population resulting from the variation in molecular content [30,58] is probably the most apparent manifestation of stochastic gene expression. In the simplest case, the concentration of some expressed protein could display some variability from cell to cell [19,39]. A more complex scenario involves populations of identical cells splitting into two or more groups, each of which is characterized by a distinct state of gene expression and growth rate [58]. Here, fluctuations in gene expression can provide the cell with a mechanism for 'sampling' physiologically distinct states, which may increase the probability of survival during times of stress without the need for genetic mutation [30,58].

5.1 Stochastic Expression of Stress-Related Genes

The probabilistic features arising from gene expression noise led to the hypothesis that evolution has fine-tuned noise-generating mechanisms and genetic architectures to derive beneficial population diversity [55,61,33]. Direct evidence that genome sequence contributes to cell-cell variability indicates that gene expression noise, like other genome-encoded traits, is inheritable and subject to selective pressures, and therefore evolvable. Specifically, large-scale proteomic studies in yeast have shown that genes associated with stress response pathways have elevated levels of intrinsic noise [6,22,36]. Stress-response genes have thus experienced positive pressure toward high population variability, presumably because this providing a selective advantage during periods of stress.

The increased gene expression noise exhibited by stress related genes lends support to the hypothesis that variability in protein content among cells might confer a selective advantage. By broadening the range of environmental stress resistance across a population, added gene expression noise could increase the likelihood that some cells within the population are better able to endure environmental assaults [5,12]. Experimental results providing support for this hypothesis were obtained in a study by Bishop et al. [9], which demonstrated a competitive advantage of stress-resistant yeast mutants under high stress due to increased phenotypic heterogeneity.

Investigations on the effect of gene expression noise have been carried out in yeast cells under acute environmental stress [10]. Both experiments and simulations confirmed that increased gene expression noise can provide a significant selective advantage at high stress levels. This was not, however, the case at low stress levels, where the low-noise strain had higher fitness than the high-noise strain.

In a qualitative explanation, Blake et al. [10] attribute the differential impact of added noise to a change in the relative fraction of surviving cells at different levels of stress. While a low-noise population will have a higher number of cells above the protein production threshold necessary for survival at low stress levels (Fig. 5a), the same will be true for a high-noise population under a high level of stress (Fig. 5b). In a quantitative model, the size of this fraction depends on the probability distribution function associated with the spread of protein content among individual cells. Consequently, if it is assumed that cells are either unaffected or killed by the stress, the population fitness (reproductive rate) and differential fitness (difference in reproductive rates between two populations, e.g. a low and a high noise cell population) for a certain stress level can be calculated (Fig. 5c and 5d, respectively) [21]. This provides a very simple quantitative framework that captures the observed impact of population heterogeneity on population fitness following acute stress.

Theoretical Models and Simulations

The impact of acute stress on the fitness W of a cell population can be calculated theoretically by evaluating the integral

$$W = \int_0^\infty w(x) f(x) dx, \tag{22}$$

where $w(x)$ is the relative reproductive rate of cells expressing a stress-related gene at a level given by x, and $f(x)$ describes the population distribution of gene expression when cells are exposed to stress [65]. In a study by Fraser et al. [21], this distribution was approximated by the lognormal distribution

$$f(x) = \frac{1}{x\beta\sqrt{2x}} \exp\left[\frac{(\ln(x) - \alpha)^2}{2\beta^2}\right], \tag{23}$$

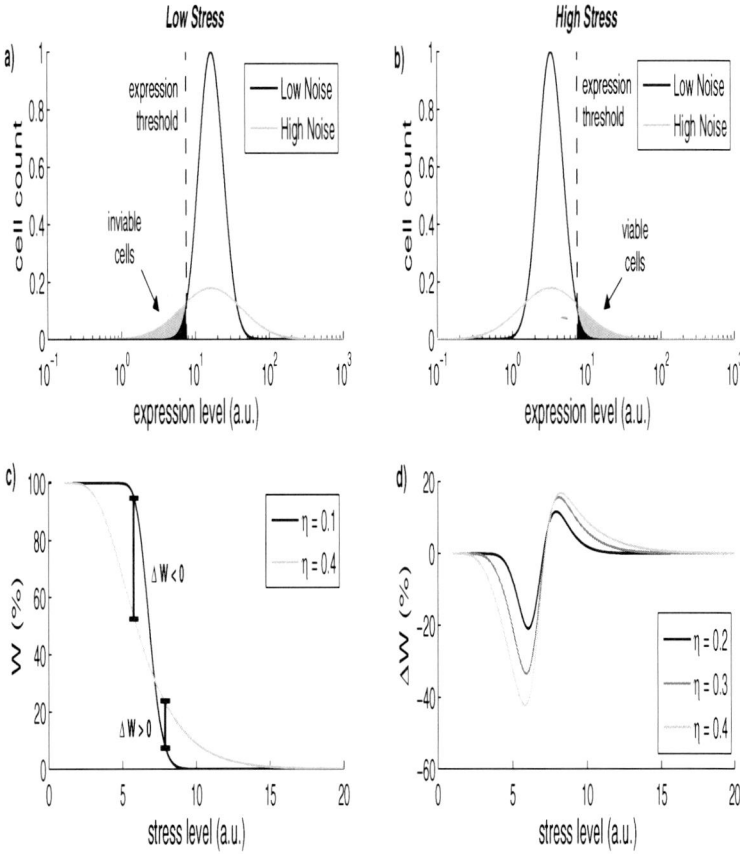

Fig. 5. Modelling the effects of noise in the expression of a stress-resistant gene. (a) Low noise is beneficial when most cells express the stress-inducing gene at levels above a certain threshold. (b) High noise is beneficial when most cells express the stress-inducing gene at levels below the threshold. (c) The effect of varying the stress level on fitness for low and high noise cell populations. Stress levels where noise is beneficial and disadvantageous are defined by positive and negative values of the differential fitness ΔW, respectively. (d) Differential fitness at varying stress levels for three populations with elevated noise relative to a low noise ($\eta_0 = 0.1$) reference population.

where α and β are defined by the average gene expression level μ and gene expression noise η through the relationships $\beta^2 = \ln(1 + \eta^2)$ and $\alpha = \ln(\mu) - 0.5\beta$. The distributions in Figure 5a and 5b were obtained for $\eta = 0.4$ and $\eta = 1.2$, respectively. Moreover, the impact of acute stress was approximated by a step function such that cells expressing a stress-resistance gene below a certain threshold would have a reproductive rate of zero, i.e., fitness $w(x) = 0$ for $x < s_{thr}$ and are otherwise unaffected, i.e., $w(x) = 1$ for $x \geq s_{thr}$.

Continuing with a positive selection scheme, where cells with high expression of a stress-resistant gene have high fitness, and cells with low expression have low fitness, we now compute W. Specifically, if it is assumed that the level of stress s experienced by the population is related to the most likely level of gene expression (i.e. the mode of the distribution in Eq. (23)), then the noise-dependency of population fitness in Eq. (22) for a threshold model is given by the error function (erf) describing the cumulative lognormal distribution

$$W(\eta, s) = \int_0^\infty w(x)f(x)dx = \int_{s_{thr}}^\infty f(x)dx$$
$$= \frac{1}{2} + \frac{1}{2}erf\left[\sqrt{\frac{\ln(1+\eta^2)}{2}}\left(\frac{\ln(s_{thr}/s)}{\ln(1+\eta^2)} - 1\right)\right]. \quad (24)$$

This equation was used to calculate the fitness curves displayed in Figure 5c using $s_{thr} = 6.91$ and $\eta = 0.1$ or $\eta = 0.4$, for the low and high noise populations respectively. Correspondingly, the differential fitness curves displayed in Figure 5d were obtained by evaluating the quantity $\Delta W(\eta, s) = W(\eta, s) - W(\eta_0, s)$, where $W(\eta, s)$ is the fitness of the population with variable high noise ($\eta = 0.2$ 0.3, or 0.4) and $W(\eta_0, s)$ is a reference population with low noise ($\eta_0 = 0.1$).

5.2 Bet-Hedging Cell Populations

Another interesting example of how noise can influence fitness involves cells that can switch between phenotypes in a changing environment [1,58]. Under fixed environmental conditions, the net growth rate (and therefore fitness) of the population is maximized when all cells are of the fastest growing phenotype. However, in a changing environment, it is thought that a statically heterogeneous population (i.e. a population where transitions between states are not influenced by environmental conditions) can deal with an uncertain future by hedging its bets. Specifically, a broad distribution of phenotypes is generated in the 'hope' that some of these phenotypes will remain viable after an environmental change. In contrast, a dynamically heterogeneous population has a more reliable strategy: individuals in such populations can sense and respond to external changes by actively switching to the fit state. If the response rate is sufficiently rapid compared to the rate of environmental fluctuations, as is the case for many real systems, then transitions from the fit state to the unfit state are actually detrimental. Thus, bet-hedging is only beneficial if response rates are sufficiently low.

Acar et al. [1] experimentally investigated how stochastic switching between phenotypes in changing environments affected growth rates in fast and slow-switching *Saccharomyces cerevisiae* (budding yeast) populations. Specifically, a strain was engineered to randomly transition between two phenotypes, ON and OFF, characterized respectively by high or low expression of a gene encoding the Ura3 enzyme, necessary for uracil biosynthesis. Each phenotype was designed to have a growth advantage over the other in one of two environments. In the first environment (E_1) uracil was lacking and cells with the ON phenotype had an advantage. In the second environment (E_2), cells with the OFF phenotype had an advantage due to the presence of a drug (5-FOA), which is converted into a toxin by the Ura3 enzyme. In this environment, which also contains uracil, cells expressing Ura3 will have low viability while cells not expressing Ura3 will grow normally.

Simulating Complex Population Dynamics

In order to simulate the scenario described above, we used a population dynamics algorithm [16] and a model of gene expression described by the following biochemical reaction scheme [30]

$$A_{act} \underset{k_2}{\overset{k_1}{\rightleftharpoons}} A_{rep} \tag{25}$$

$$A_{act} \xrightarrow{s_{A,act}} A_{act} + M \tag{26}$$

$$A_{rep} \xrightarrow{s_{A,rep}} A_{rep} + M \tag{27}$$

$$M \xrightarrow{s_P} M + P \tag{28}$$

$$M \xrightarrow{\delta_M} \oslash \tag{29}$$

$$P \xrightarrow{\delta_P} \oslash \tag{30}$$

Eq. (25) describes the transitions to the active (upregulated level of gene expression) A_{act} and repressed (basal level of gene expression) A_{rep} promoter states with rates k_1 and k_2 respectively, Eqs. (26) and (27) the mRNA production from the A_{act} (at a rate $s_{A,act}$) and A_{rep} (at a rate $s_{A,rep}$) states respectively, Eq. (29) the protein production from mRNA at a rate s_P, and Eqs. (28) and (30) respectively the mRNA (at a rate δ_M) and protein (at a rate δ_P) degradation. The fitness w_k of each cell k, which is here defined as a function of the environment and cellular protein concentration $[P]$, was described by a Hill function

$$w_k(E, [P]) = \begin{cases} \frac{[P]^n}{[P]^n + K^n}, & if\ E = E1, \\ \frac{K^n}{K^n + [P]^n}, & if\ E = E2. \end{cases} \tag{31}$$

This equation describes partitioning of cells into fit ($w_k(E, P) \geq 0.5$) and unfit ($w_k(E, P) < 0.5$) phenotypes corresponding to whether or not their $[P]$ in a particular environment is above or below a particular value given by the Hill coefficient K. The volume of each cell was modelled using an exponential growth law

$$V_k(t_{div}) = V_0 \exp\left[\ln(2)\left(\frac{t_{div}}{\tau_0}\right)\right]. \tag{32}$$

Here, V_0 is the cell volume at the time of its birth, and $\tau_0 = \tau_\phi/w$, where τ_ϕ is the cell division time in absence of any selective pressure. To incorporate the effect of fitness on gene expression, the value of transcription rate parameter s_A depended on whether or not a cell was fit in either $E1$ or $E2$ (see Fig. 6 and [1] for parameters). Note that in this model the cells divided symmetrically when their volume reached $2V_0$.

The population distributions obtained for this model are shown in Figure 6. Specifically, we first obtained the steady-state protein concentration distributions for cells in $E1$ and $E2$ (Fig. 6a and 6b, respectively). Here, the majority of cells either fell within a distribution centered at higher value of P, characterizing the ON cells, or a distribution centered at a lower value, characterizing the OFF cells, in $E1$ or $E2$ respectively. The rest of the cells fell within the distribution capturing the unfit subpopulation in both environments. These results were found experimentally in [1] and are expected, as higher levels of the Ura3 enzyme are either favorable or unfavorable with respect to the fitness of

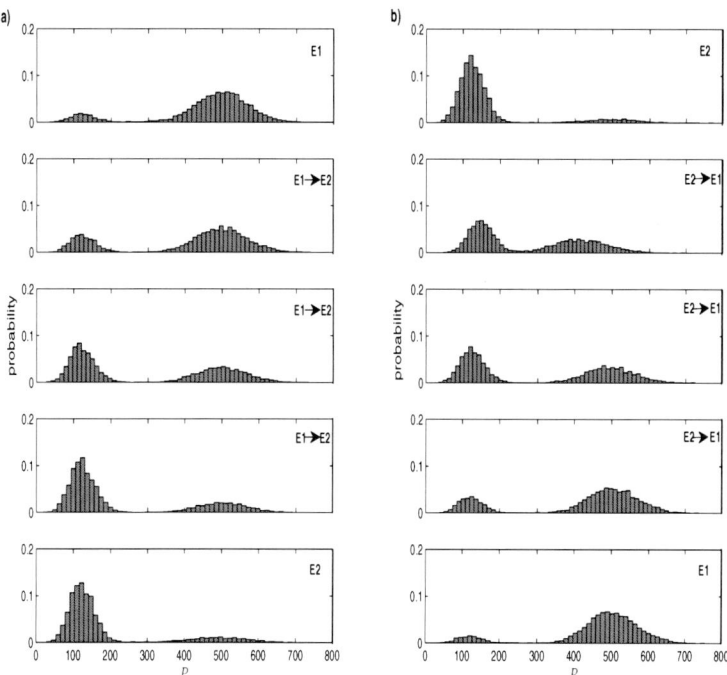

Fig. 6. Simulations of environmental effects on phenotypic distribution. (a) Steady-state (top and bottom figures) and time-dependent (middle figures) protein distributions of cells transfered from an environment lacking uracil (E1) to an environment containing uracil and 5-FOA (E2). (b) Steady-state (top and bottom figures) and time-dependent (middle figures) protein distributions of cells transfered from E1 to E2. Note that when a sufficient amount of time has elapsed after the environmental transition from either E1 to E2 or vice versa, cells with either the OFF or ON phenotype proliferate, respectively, in agreement with experimental results found in [1]. The following parameters were used (units s^{-1}): $\delta_M = 0.005$, $s_P = 0.1$, $\delta_P = 0.008$, $K = 200$, $n = 10$. For fit cells in E1 $s_{A,act} = 0.2$ and for unfit cells $s_{A,rep} = 0.05$ - vice versa in E2. Additionally τ_ϕ was set to the mean doubling time (MDT) of 1.5 hours for *Saccharomyces cerevisiae* [13].

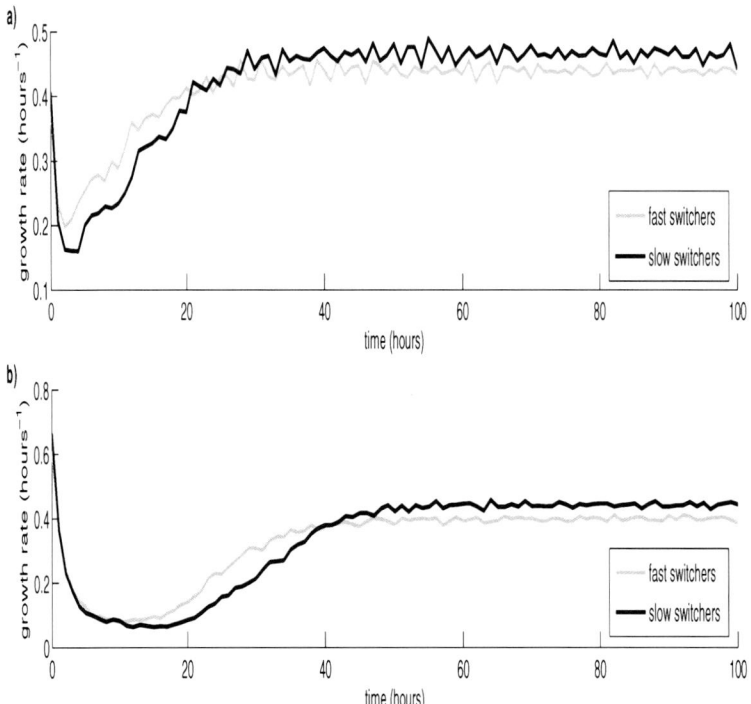

Fig. 7. Simulations of populations of slow and fast-switching cells. (a) Growth rates of cells after an environmental change from E2 to E1 at $t = 0$. (b) Growth rates of cells after environmental change from E1 to E2 at $t = 0$. Note that the transient before the steady-state region is shorter in (a) than in (b), and that fast-switching cells recover faster from the environment change but slow-switching cells have a higher steady-state growth, in agreement with experimental results found in [1].

the cells depending on the environment. Additionally, the time-dependent population distributions after the transition to $E1$ from $E2$, and vice versa, were obtained (Fig. 6a and 6b, respectively). Here, the dynamics of the two distinct subpopulations of cells in transition between the steady-states are visible. As time progresses after the environmental transition, fewer and fewer of the cells are in the unfit state (ON in Fig. 6a and OFF in Fig. 6b), as the cells in the more fit state (OFF in Fig. 6a and ON in Fig. 6b) grow and divide at a faster rate and therefore come to dominate the population in terms of absolute numbers. Figure 7 shows the growth rates obtained from simulations of slow and fast-switching cell populations, where cells were transfered from E2 to E1, and vice versa, at $t = 0$. Growth rates show a transition period and a steady-state region. In agreement with experiments (see Acar *et al.* [1]), fast-switching cells were found to recover from the effect of environment change faster than slow-switching cells but have a lower steady-state growth rate.

6 Cellular Decision-Making in a Noisy Environment

Previous sections have described sources of noise in gene regulatory networks, how noise can impact fitness, and how different regulatory mechanisms can either attenuate or amplify noise. While noise is an inherent part of the stochastic chemistry of cells, it is also an inherent part of their sensing apparatus as well as of the signals they sense. For example, cells can respond to the concentrations of numerous kinds of chemicals, including nutrients, toxins, signaling molecules, as well as physical properties of the environment such as pressure and temperature. A recent line of research has investigated models of how cells should process such noisy signals, and in particular, whether human theories of optimal signal processing might be embodied in cells–implemented chemically, as it were [17,2,3,32]. We present a simplified version of the analysis of Libby *et al.* [32]. We show that it is possible, in principle, for the chemistry of gene regulation to approximate probability-theoretic computations related to the analysis of noisy signals. This general viewpoint provides one possible interpretation, a detailed quantitative interpretation, for the function of real regulatory networks.

6.1 Two-Class Bayesian Discrimination Problems

The work of Libby *et al.* [32] used the framework of two-class Bayesian discrimination problems to interpret gene regulatory mechanisms, and the lac operon of *E. coli* in particular. In these problems, we imagine that there is an unobserved binary random variable X, whose value one wants to estimate. For example, it may be important to an *E. coli* whether its immediate environment has a low or high concentration of a particular sugar (Figure 8), in order for it to make the right choices about expressing genes useful for the import and metabolism of that sugar. In other cases, the relevant variable may be the presence or absence

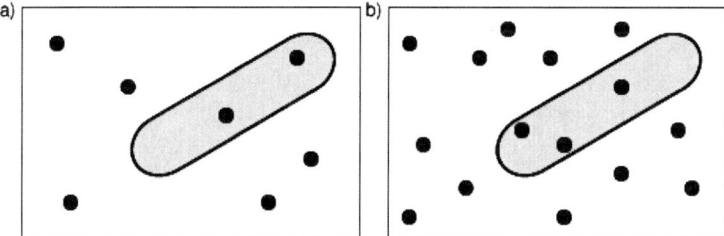

Fig. 8. Conceptualization of an inference problem solved by a cell. (a) An *E. coli* cell (oblong) in an environment low in a particular sugar (black circles). (b) The same cell in a higher sugar environment. The amount of intracellular sugar is related, albeit imperfectly and stochastically, to the extracellular sugar concentration. While intracellular sugar directly drives the regulation of genes related to its metabolism, it is the extracellular sugar that is of true importance to the regulatory decisions made by the cell.

of a toxin, a mating partner, a competitor organism, etc. Although X is not directly observed, we assume there is another variable S which is observed, and the value of which depends stochastically on the value of X. For example, in the situation depicted in Figure 8, S may be the intracellular sugar concentration. This S can be viewed as "observed by" or "known to" the cell, because this sugar can interact directly, chemically, with the regulatory machinery of the cell and bring about changes in cellular behavior (e.g., changing the expression of certain genes). The exact value of S may depend on many factors—the size of the cell, the number of permeases, and so on, but it clearly depends as well on the extracellular environment state, X. We can imagine that there are different probability distributions for S depending on the state X, $P(S = s|X = low)$ an $P(S = s|X = high)$. The problem the cell faces, then, is to estimate the probabilities of $X = low$ and $X = high$ based on the signal value $S = s$. This can be done via Bayes's rule

$$P(X = high|S = s) = \frac{P(S = s|X = high)P(X = high)}{P(S = s)}$$

$$= \frac{P(S = s|X = high)P(X = high)}{P(S = s|X = high)P(X = high) + P(S = s|X = low)P(X = low)}.$$

From this formula, it is clear that the probability of X being high or low depends not just on the value of S, via the probability distribution for S as a function of X, but also on the terms $P(X = high)$ and $P(X = low) = 1 - P(X = high)$. The are called the prior probabilities, which are one's beliefs about X before the signal S has been accounted for, while $P(X = high|S = s)$ and $P(X = low|S = s) = 1 - P(X = high|S = s)$ are called the posterior probabilities, representing one's beliefs about X after the signal S has been accounted for.

6.2 A Model of Genetic Response to Intracellular Sugar

We present a simplified chemical model of gene activation that is broadly similar to the function of the lac operon of *E. coli*, as well as a number of other sugar metabolic systems. It is not intended as a description of the lac operon per se, but rather as a generic model of negatively regulated control. We model intracellular sugar, S, a repressor molecule R, and the promoter A of a gene whose protein P is expressed in a correlated fashion to sugar S. In a real system, P might actually represent a set of proteins involved in the metabolism or import of the sugar S, but we do not model these aspects. We merely think of P as being the response of the cell that is turned on by the presence of S.

$$\emptyset \xrightarrow{r_S(X)} S \tag{33}$$

$$S \xrightarrow{\gamma_S} \emptyset \tag{34}$$

$$S + R \underset{r_{SR}}{\overset{r_{RS}}{\rightleftharpoons}} SR \tag{35}$$

$$SR \xrightarrow{\gamma_S} R \tag{36}$$

$$A + R \underset{r_{RA}}{\overset{r_{AR}}{\rightleftharpoons}} AR \tag{37}$$

$$A \xrightarrow{s_P} A + P \tag{38}$$

$$P \xrightarrow{\delta_P} \oslash \tag{39}$$

Eq. (33) describes the process of intracellular sugar entering the system at rate $r_S(X)$, which, because X is binary, can be one of two values—r_{low} when $X = low$ and r_{high} when $X = high$. Sugar "decays", whether bound to the repressor (Eq. (36)) or not (Eq. (34)), which would realistically represent the sugar being metabolized, or concentration decreasing via dilution. The repressor can bind to the promoter and make it transcriptionally inactive (Eq. (37)). However, a repressor molecule bound by sugar (Eq. 35) cannot bind the promoter. In this way, increasing S leads to decreasing free R, and thus increasing transcriptional activation and increasing level of P. At this qualitative level, the model behaves as would be expected by a system that responds to the sugar S. Is a more quantitative interpretation of the system possible? Is it possible for the system to implement, or approximate, the Bayesian two-class computation described above, so that the "output" of the system, the expression of the protein P, is proportional to the posterior probability that the external environment being in state $X = high$?

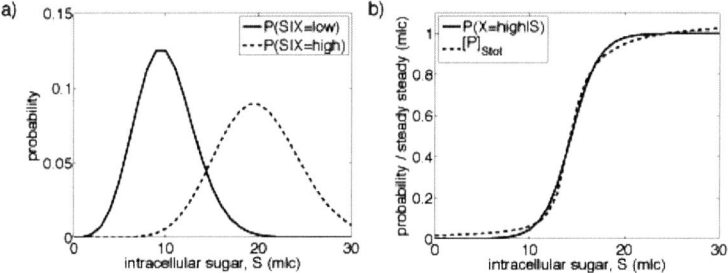

Fig. 9. Conditional and posterior probabilities for a problem of inferring environment state X (low or high in sugar) based on the noisy intracellular sugar level, S. (a) The probability distributions for S in the two environment states. (b) The posterior probability of $X = high$ given an intracellular sugar level S, and the output of the chemical model of gene regulation, with parameters tuned to match the posterior probability.

6.3 Chemically Approximating Bayesian Two-Class Discrimination

Libby et al. [32] showed that a variety of different chemical regulatory models of sugar metabolism are indeed capable of approximating the Bayesian two-class computation. To demonstrate this using the model above, suppose that $r_{low} = 10$ molecules per second, $r_{high} = 20$ molecules per second, and $\gamma_S = 1 \ s^{-1}$. When $X = low$, the steady-state probability distribution for S is Poisson with parameter $\lambda = 10$, and when $X = high$, it is Poisson with parameter $\lambda = 20$ (Figure 9a). Assuming that $X = low$ and $X = high$ are equally likely a priori, so that $P(X = low) = P(X = high) = \frac{1}{2}$, then Equation 6.1 can be used to compute the posterior probability that the environment is in a high sugar state. The result of this computation is shown in Figure 9b.

Returning to the chemical model, let $[P]_{S_{tot}}$ denote the steady-state number of molecules of P when the total intracellular sugar $S_{tot} = S + RS$ is fixed at a certain level. That is, we remove reactions 33, 34 and 36 from the model, and compute the (deterministic) steady-state of the system. We implemented this steady-state computation in `Matlab` and used the `fminsearch` utility to find reaction rate parameters for the system that minimize the squared error function

$$\sum_{S_{tot}=0}^{30} (P(X = high|S_{tot}) - [P]_{S_{tot}})^2. \tag{40}$$

As shown in Figure 9b, the parameters of the chemical model can be chosen so that the average number of molecules of P, given intracellular sugar level S_{tot}, closely matches the Bayesian computation of the probability that the environment is in the high sugar state. This demonstrates that even the simplest gene regulatory mechanisms are capable, in principle, of approximately reproducing fairly sophisticated probability-theoretic computations, and thus are capable of implementing inferential procedures to help the cell reason about its environment.

Whether or not this is an appropriate interpretation of the behavior of real gene regulatory systems remains to be seen. Libby et al. [32] showed that the experimentally measured response of the lac operon to two signals, lactose concentration and cAMP concentration (a starvation signal), is consistent with a solution to a two-class discrimination problem. Relatedly, Dekel et al. [17] showed that expression of the lac operon seems to balance the metabolic benefit from the sugar against the metabolic cost of expression. Andrews et al. [2,3] have shown that chemotactic behavior can be interpreted through the lens of filtering and information theory. Thus, there is growing evidence that human theories of noisy signal processing and decision making may indeed be implemented biochemically in the cell, and that these theories provide explanations for the detailed quantitative behaviors of cellular networks.

7 Conclusion

Our understanding of the origins and consequences of stochasticity in gene expression has advanced significantly in recent years. This advancement has been fueled by theoretical developments enabling biological hypothesis formulation using stochastic process and dynamical systems theory, as well as experimental breakthroughs in measurements of gene expression at the single cell level [53].

Noise in gene expression was originally viewed as being detrimental in terms of cellular function due to the corruption of intracellular signals negatively affecting cellular regulation with possible implications for disease. However, noisy gene expression can also be advantageous, providing the flexibility needed by cells to adapt to stress such as a changing environment [1,21,58]. Stochasticity in gene expression provides a mechanism for the occurrence of heterogeneous populations of genetically identical cells, in terms of phenotypic and cell-type diversity, which can be established during cellular growth and division [14,30,51]. Furthermore, studies have suggested that intrinsic stochasticity in gene expression is an evolvable trait [22,39].

Gene expression noise not only arises from intrinsic fluctuations, but also from noise propagated through the network from upstream genes [42]. Several genetic network motifs including cascades and feedback loops can act to modulate this noise, resulting in a range of behaviour including amplification, bounded fluctuations, and noise filtration [42,49,59].

Cells depend on the information they obtain from their environment to remain viable. Yet this information, received at the cell surface, is conveyed through gene and protein networks and is transferred via biochemical reactions that are inherently stochastic [11,19,39,45]. Stochastic fluctuations can undermine both signal detection and transduction. As a result, cells are confronted with the task of predicting the state of the extracellular environment from noisy and potentially unreliable intracellular signals. In addition to employing noise reduction mechanisms, cells may statistically infer the state of the extracellular environment from intracellular inputs [32,43].

The study of noise in genetic networks has provided novel insights into how cells survive, propagate and ultimately perish in stochastic environments. This line of research is likely to continue to prove fundamental for developments in the fields of molecular and synthetic biology and in furthering our understanding and treatment of human disease.

Acknowledgments

The authors would like to thank Hilary Phenix for editing the chapter and the National Science and Engineering Research Council of Canada for research funding.

References

1. Acar, M., Mettetal, J.T., van Oudenaarden, A.: Stochastic switching as a survival strategy in fluctuating environments. Nat. Genet. 40, 471–475 (2008)
2. Andrews, B.W., Yi, T.-M., Iglesias, P.A.: Optimal Noise Filtering in the Chemotactic Response of *Escherichia coli*. PLoS Comput. Biol. 2(11), 154 (2006)

3. Andrews, B.W., Iglesias, P.A.: An Information-Theoretic Characterization of the Optimal Gradient Sensing Response of Cells. PLoS Comput. Biol. 3(8), e153 (2007)
4. Austin, D.W., Allen, M.S., McCollum, J.M., Dar, R.D., et al.: Gene network shaping of inherent noise spectra. Nature 439, 608–611 (2006)
5. Avery, S.V.: Microbial cell individuality and the underlying sources of heterogeneity. Nat. Rev. Microbiol. 4, 577–587 (2006)
6. Bar-Even, A., Paulsson, J., Maheshri, N., Carmi, M., et al.: Noise in protein expression scales with natural protein abundance. Nat. Genet. 38, 636–643 (2006)
7. Bar-Or, R.L., Maya, R., Segel, L.A., Alon, U., et al.: Generation of oscillations by the p53-Mdm2 feedback loop.: A theoretical and experimental study. PNAS 97, 11250–11255 (2000)
8. Becskei, A., Serrano, L.: Engineering stability in gene networks by autoregulation. Nature 405, 590–593 (2000)
9. Bishop, A.L., Rab, F.A., Sumner, E.R., Avery, S.V.: Phenotypic heterogeneity can enhance rare-cell survival in 'stress-sensitive' yeast populations. Molec. Microbiol. 63, 507–520 (2007)
10. Blake, W.J., Balazsi, G., Kohanski, M.A., Isaacs, F.J., et al.: Phenotypic consequences of promoter-mediated transcriptional noise. Mol. Cell 24, 853–865 (2006)
11. Blake, W.J., Kaern, M., Cantor, C.R., Collins, J.J.: Noise in eukaryotic gene expression. Nature 422, 633–637 (2003)
12. Booth, I.R.: Stress and the single cell: intrapopulation diversity is a mechanism to ensure survival upon exposure to stress. Int. J. Food Microbiol. 78, 19–30 (2002)
13. Brewer, B.J., Chlebowicz-Sledziewska, E., Fangman, W.L.: Cell Cycle Phases in the Unequal Mother/Daughter Cell Cycles of *Saccharomyces cerevisiae*. Mol. Cell. Biol. 4, 2529–2531 (1984)
14. Brock, A., Chang, H., Huang, S.: Non-genetic heterogeneity - a mutation-independent driving force for the somatic evolution of tumours. Nat. Rev. Genet. 10, 336–342 (2009)
15. Chang, H.H., Hemberg, M., Barahona, M., Ingber, E., Huang, S.: Transcriptome-wide noise controls lineage choice in mammalian progenitor cells. Nature 453, 544–547 (2008)
16. Charlebois, D.A., Intosalmi, J., Fraser, D., Kaern, M.: An Algorithm for the Stochastic Simulation of Gene Expression and Heterogeneous Population Dynamics. Commun. Comput. Phys. 9, 89–112 (2011)
17. Dekel, E., Alon, U.: Optimality and evolutionary tuning of the expression level of a protein. Nature 436, 588–592 (2005)
18. Dublanche, Y., Michalodimitrakis, K., Kümmerer, N., Foglierini, M., et al.: Noise in transcription negative feedback loops: simulation and experimental analysis. Molec. Syst. Biol. (2006), doi:10.1038/msb4100081
19. Elowitz, M.B., Levine, A.J., Siggia, E.D., Swain, P.S.: Stochastic gene expression in a single cell. Science 297, 1183–1186 (2002)
20. Ferrell Jr., J.E., Machleder, E.M.: The Biochemical Basis of an All-or-None Cell Fate Switch in *Xenopus Oocytes*. Science 8, 895–898 (1998)
21. Fraser, D., Kaern, M.: A chance at survival: gene expression noise and phenotypic diversification strategies. Molec. Microbiol. 71, 1333–1340 (2009)
22. Fraser, H.B., Hirsh, A.E., Giaever, G., Kumm, J., et al.: Noise minimization in eukaryotic gene expression. PLoS Biol. 2, e137 (2004)
23. Gambacorti-Passerini, C.B., Gunby, R.H., Piazza, R., Galietta, A., et al.: Molecular mechanisms of resistance to imatinib in Philadelphia-chromosome-positive leukaemias. Lancet. Oncol. 4, 75–85 (2003)

24. Gillespie, D.T.: A general method for numerically simulating the stochastic time evolution of coupled chemical reactions. J. Comput. Phys. 22, 403–434 (1976)
25. Gillespie, D.T.: Exact stochastic simulation of coupled chemical reactions. J. Phys. Chem. 81, 2340–2361 (1977)
26. Gillespie, D.T.: Stochastic Chemical Kinetics. In: Yip, S. (ed.) Handbook of Materials and Modeling, sec. 5.11. Springer, Heidelberg (2005)
27. Gorre, M.E., Mohammed, M., Ellwood, K., Hsu, N., et al.: Clinical resistance to STI-571 cancer therapy caused by BCR-ABL gene mutation or amplification. Science 293, 876–880 (2001)
28. Hooshangi, S., Thiberge, S., Weiss, R.: Ultrasensitivity and noise propagation in a synthetic transcriptional cascade. PNAS 102, 3581–3586 (2005)
29. Hornung, G., Barkai, N.: Noise Propagation and Signaling Sensitivity in Biological Networks: A Role for Positive Feedback. PLoS Comput. Biol. 4 (2008), doi:10.1371/journal.pcbi.0040008
30. Kaern, M., Elston, T.C., Blake, W.J., Collins, J.J.: Stochasticity in gene expression: From theories to phenotypes. Nat. Rev. Genet. 6, 451–464 (2005)
31. Kaufmann, B.B., van Oudenaarden, A.: Stochastic gene expression: from single molecules to the proteome. Curr. Opin. Genet. Dev. 17, 107–112 (2007)
32. Libby, E., Perkins, T.J., Swain, P.S.: Noisy information processing through transcriptional regulation. PNAS 104, 7151–7156 (2007)
33. Lopez-Maury, L., Marguerat, S., Bahler, J.: Tuning gene expression to changing environments: from rapid response to evolutionary adaptation. Nat. Rev. Gen. 9, 583–593 (2008)
34. Ma, L., Wagner, J., Rice, J.J., Wenwei, H., Arnold, J.L., Stolovitzky, G.A.: A plausible model for the digital response of p53 to DNA damage. PNAS 102, 14266–14271 (2005)
35. Maheshri, N., O'Shea, E.K.: Living with noisy genes: how cells function reliably with inherent variability in gene expression. Annu. Rev. Biophys. Biomol. Struct. 36, 413–434 (2007)
36. Newman, J.R.S., Ghaemmaghami, S., Ihmels, J., Breslow, D.K., et al.: Single-cell proteomic analysis of S. *cerevisiae* reveals the architecture of biological noise. Nature 441, 840–846 (2006)
37. Okabe, S., Tauchi, T., Ohyashiki, K.: Characteristics of dasatinib- and imatinib-resitant chronic myelogenous leukemia cells. Clin. Cancer Res. 14, 6181–6186 (2008)
38. Orrell, D., Bolouri, H.: Control of internal and external noise in genetic regulatory networks. J. Theor. Biol. 230, 301–312 (2004)
39. Ozbudak, E.M., Thattai, M., Kurtser, I., Grossman, A.D., et al.: Regulation of noise in the expression of a single gene. Nat. Genet. 31, 69–73 (2002)
40. Paulsson, J.: Noise in a minimal regulatory network: plasmid copy number control. Quart. Rev. Biophys. 34, 1–59 (2001)
41. Paulsson, J.: Summing up the noise in gene networks. Nature 427, 415–418 (2004)
42. Pedraza, J.M., van Oudenaarden, A.: Noise propagation in gene networks. Science 307, 1965–1969 (2005)
43. Perkins, T.J., Swain, P.S.: Strategies for cellular decision-making. Molec. Syst. Biol. (2009), doi:10.1038/msb.2009.83
44. Raj, A., van Oudenaarden, A.: Nature, Nurture, or Chance: Stochastic Gene Expression and Its Consequences. Cell (2008), doi:10.1016/j.cell.2008.09.050
45. Raser, J.M., O'Shea, E.K.: Control of stochasticity in eukaryotic gene expression. Science 304, 1811–1814 (2004)
46. Raser, J.M., O'Shea, E.K.: Noise in gene expression: origins, consequences, and control. Science 309, 2010–2013 (2005)

47. Ribeiro, A.S., Zhu, R., Kauffman, S.A.: A General Modeling Strategy for Gene Regulatory Networks with Stochastic Dynamics. J. Comput. Biol. 13, 1630–1639 (2006)
48. Rosenfeld, N., Elowitz, M.B., Alon, U.: Negative autoregulation speeds the response times of transcription networks. J. Mol. Biol. 323, 785–793 (2002)
49. Rosenfeld, N., Young, J.W., Alon, U., Swain, P.S., Elowitz, M.B.: Gene regulation at the single-cell level. Science 307, 1962–1965 (2005)
50. Roussel, M., Zhu, R.: Validation of an algorithm for the delay stochastic simulation of transcription and translation in prokaryotic gene expression. Phys. Biol. 3, 274–284 (2006)
51. Samoilov, M.S., Price, G., Arkin, A.P.: From fluctuations to Phenotypes: The Physiology of Noise. Sci. STKE 366, 17 (2006)
52. Savageau, M.A.: Comparison of classical and autogenous systems of regulation in inducible operons. Nature 252, 546–549 (1974)
53. Scott, M., Ingalls, B., Kaern, M.: Estimations of intrinsic and extrinsic noise in models of nonlinear genetic networks. Chaos 16 026107-1–026107-15 (2006)
54. Simpson, M.L., Cox, C.D., Sayler, G.S.: Frequency domain analysis of noise in autoregulated gene circuits. PNAS 100, 4551–4556 (2003)
55. Smits, W.K., Kuipers, O.P., Veening, J.W.: Phenotypic variation in bacteria: the role of feedback regulation. Nat. Rev. Microbiol. 4, 259–271 (2006)
56. Swain, P.S., Elowitz, M.B., Siggia, E.D.: Intrinsic and extrinsic contributions to stochasticity in gene expression. PNAS 99, 12795–12800 (2002)
57. Thattai, M., van Oudenaarden, A.: Intrinsic noise in gene regulatory networks. PNAS 98, 8614–8619 (2001)
58. Thattai, M., van Oudenaarden, A.: Stochastic Gene Expression in Fluctuating Environments. Genetics 167, 523–530 (2004)
59. Thattai, M., van Oudenaarden, A.: Attenuation of noise in ultrasensitive signaling cascades. Biophys. J. 82, 2943–2950 (2002)
60. Turner, T.E., Schnell, S., Burrage, K.: Stochastic approaches for modelling in vivo reactions. Comput. Biol. Chem. 28, 165–178 (2004)
61. Veening, J.W., Smits, W.K., Kuipers, O.P.: Bistability, epigenetics, and bet-hedging in bacteria. Annu. Rev. Microbiol. 62, 193–210 (2008)
62. Volfson, D., Marciniak, J., Blake, W.J., Ostroff, N., et al.: Origins of extrinsic variability in eukaryotic gene expression. Nature 439, 861–864 (2006)
63. Wei, Y., Hardling, M., Olsson, B., Hezaveh, R., et al.: Not all imatinib resistance in CML are BCR-ABL kinase domain mutations. Ann. Hematol. 85, 841–847 (2006)
64. Wilkinson, D.J.: Stochastic Modelling for Systems Biology. Chapman and Hall, Boca Raton (2006)
65. Zhuravel, D., Fraser, D., St-Pierre, S., Tepliakova, L., Pang, W.L., Hasty, J., Kaern, M.: Phenotypic impact of regulatory noise in cellular stress-response pathways. Syst. Synth. Biol. (2010), doi:10.1007/s11693-010-9055-2

Boolean Threshold Networks: Virtues and Limitations for Biological Modeling

Jorge G.T. Zañudo, Maximino Aldana, and Gustavo Martínez-Mekler

Instituto de Ciencias Físicas, Universidad Nacional Autónoma de México,
Avenida Universidad s/n, Colonia Chamilpa, Cuernavaca, Morelos, México,
Código Postal 62210
max@fis.unam.mx
http://www.fis.unam.mx/~max/

Abstract. Boolean threshold networks have recently been proposed as useful tools to model the dynamics of genetic regulatory networks, and have been successfully applied to describe the cell cycles of *S. cerevisiae* and *S. pombe*. Threshold networks assume that gene regulation processes are additive. This, however, contrasts with the mechanism proposed by S. Kauffman in which each of the logic functions must be carefully constructed to accurately take into account the combinatorial nature of gene regulation. While Kauffman Boolean networks have been extensively studied and proved to have the necessary properties required for modeling the fundamental characteristics of genetic regulatory networks, not much is known about the essential properties of threshold networks. Here we study the dynamical properties of these networks with different connectivities, activator-repressor proportions, activator-repressor strengths and different thresholds. Special attention is paid to the way in which the threshold value affects the dynamical regime in which the network operates and the structure of the attractor landscape. We find that only for a very restricted set of parameters, these networks show dynamical properties consistent with what is observed in biological systems. The virtues of these properties and the possible problems related with the restrictions are discussed and related to earlier work that uses these kind of models.

1 Introduction

The analysis of the dynamics of genetic regulatory networks in living organisms is a complicated task and a central challenge in current research for a complete understanding of complex biological systems. Historically, the dynamical behaviour of the biochemical elements in small genetic circuits has been accurately described using differential equations, which capture the underlying reaction-diffusion kinetics that take place in these systems [1,2,3]. However, this approach faces important difficulties for the modeling of large genetic networks, being the main difficulty that these mathematical models may involve a very large amount of parameters. This is a serious problem both practically and theoretically. Practically since these parameters may be largely unknown for many systems, and theoretically since such a detailed description may obscure the essential properties of the regulatory processes in the systems under consideration[4,5]. Because of this, Boolean networks have recently been increasingly used as the best first approach for the modeling and understanding of the essential properties of real regulatory systems that incorporate large amounts of data [6,7].

Boolean networks have been extensively studied for decades [8], and were introduced for the modeling of large regulatory systems by S. Kauffman as a first attempt to understand the general dynamical properties of the gene regulation and cell differentiation processes [9,10]. However, it was only recently that the necessary information to test them on real biological genetic networks has been available. Examples are models of the genetic network of flower development in *Arabidopsis thaliana* [11,12], the regulatory network determining embryonic segmentation in *Drosophila melanogaster* [13], the network controlling the differentiation process in Th cells [14], the cell cycle networks of *Saccharomyces cerevisiae* [15] and *Saccharomyces pombe* [16], among others. One of the advantages of the Boolean approach is that it is not necessary to know the kinetic details of the interactions (e.g. promoter affinities, degradation constants, translation rates, etc.). Rather, only the logic of the regulatory interactions is needed, such as the specific activatory or inhibitory nature of the genetic regulations [4,5]. By incorporating this information, available nowadays from high-throughput experiments, into the Boolean approach, it has been possible to predict the temporal sequence of gene activities as well as the stable and periodic patterns of gene expression in wild type and in many mutants of the organisms mentioned before.

There are, however, important differences in the way in which Boolean models have been implemented by different groups, and it is not clear whether or not these different implementations would yield equivalent results. The general formulation of the Boolean network model is the following. We assume that the network is represented by a set of N Boolean variables (or genes) $\{\sigma_1, \sigma_2, \ldots, \sigma_N\}$, each of which can be in two different states $\sigma_i = 1$ (active) and $\sigma_i = 0$ (inactive). The state of each gene σ_i is controlled by k_i other genes of the network, $\{\sigma_{i_1}, \ldots, \sigma_{i_{k_i}}\}$, which we will refer to as the *regulators* or the *inputs* of σ_i. The number k_i of regulators of each gene depends on the topology of the network in such a way that the probability for a randomly selected node to have k regulators is given by the probability distribution $P_{in}(k)$. Once every gene has been provided with a set of regulators, the dynamics of the network are given by the simultaneous updating of all the gene states according to

$$\sigma_i(t+1) = F_i\left(\sigma_{i_1}(t), \sigma_{i_2}(t), \ldots, \sigma_{i_{k_i}}(t)\right), \qquad (1)$$

where F_i is a regulatory function, specific to the gene σ_i, that is constructed according to the activatory and inhibitory nature of the regulators of σ_i.

The differences in the implementation of the Boolean approach mentioned before are related to the way in which the regulatory functions F_i are constructed.[1] For instance, the regulatory functions used in [11,12] for the *A. thaliana* flower development network were carefully constructed taking into account the current biological knowledge about the *combinatorial* action of the regulators on their target genes. This combinatorial action takes place, for instance, with dual regulators whose inhibitory or activatory nature on the target gene depends upon the presence or absence of other regulators (which may compete for the same binding site in the promoter region) [17]. In contrasts, the regulatory functions used in [15] and [16] for the cell cycle networks of *S. cerevisiae* and *S. pombe* are threshold functions similar to the ones used in artifical neural networks [18,19]. These two schemes, combinatorial functions vs. threshold functions, are very different not only mathematically, but in their very nature. For the use of threshold functions requires the strong assumption that *the effect of activatory and inhibitory regulations, rather than combinatorial, is simply additive.*

In spite of this strong assumption, Boolean models with threshold functions seem to predict the correct biological sequence of events in the cell cycles of *S. cerevisiae* and *S. pombe* [15,16]. The dynamics in each of these systems exhibit one big attractor that corresponds to the experimentally observed stable state at the end of the cell cycle. This result suggests that, under certain conditions, gene regulatory interactions can indeed be considered as purely additive. In such cases, Boolean models with threshold functions are useful to describe real genetic networks and understand their dynamical properties [20]. Therefore, a thorough study of these kind of mathematical models is necessary. However, although Boolean networks with threshold functions have been extensively studied in the context of spin glasses [21]-[26] and artificial neural networks [18,19], their dynamical properties in the context of gene regulation are largely unknown. For only the most simple cases of fixed connectivities, equal activator/repressor sterngths and proportions, and fixed threshold values have been explored [20,27].[2]

In this work we investigate the generic dynamical properties of Boolean networks with threshold functions. Our main goal is to compare the behavior of these threshold networks with the one that is already known for standard random Boolean networks (also termed Kauffman networks), focusing on the properties that are relevant to gene regulation processes. To this end, we use different connectivities, activator/repressor strengths and proportions, and threshold values. In Sec. 2 we describe the Boolean threshold network model and present

---

[1] There is another important difference in the Boolean implementation which is not related to the regulatory functions but that is worth mentioning, which is the synchronous versus the asynchronous updating schemes. Throughout this work we will use synchronous updating because we want to focus on the differences regarding the construction of the regulatory functions.

[2] Usually, for spin glasses and neural networks the nodes σ_i take the values $\{+1, -1\}$ (rather than $\{1, 0\}$). Although models using the spin-like values $\{+1, -1\}$ can be mapped onto models using $\{1, 0\}$, the mapping requires a fine tuning of the threshold values θ_i.

examples of strong deviations from the "normal" behavior observed in Kauffman Boolean networks. Next, in Sec. 3 we use the annealed approximation [28] and the average influence [29] of these networks to calculate the phase diagram for the different parameters involved. In Sec. 4 we present numerical evidence to support the analytical results and discuss the case where anomalies between the theoretical prediction and the numerical simulations arise. Finally, we discuss and summarize our results, highlighting their implications in terms of the applicability of threshold networks for the modeling of gene regulation.

2 The Boolean Threshold Network Model

2.1 Definition and General Properties

In what follows we will refer to Boolean networks with threshold functions as *Boolean threshold networks* (or BTN's). Since threshold functions are a subset of the general class $\mathcal{B}$ comprising all possible Boolean functions, it is clear that BTN's are a subset of the ensemble of random Boolean networks (RBN's) introduced by Kauffman, in which the regulatory functions F_i are randomly chosen from $\mathcal{B}$. In the context of gene regulation, the dynamics of BTN's are given by

$$\sigma_i(t+1) = F_i\left(\sigma_{i_1}(t), \ldots, \sigma_i(t), \ldots, \sigma_{i_{k_i}}(t)\right) = \begin{cases} 1, & \sum_{j=1}^{k_i} a_{i,j}\sigma_{i_j}(t) > \theta_i \\ 0, & \sum_{j=1}^{k_i} a_{i,j}\sigma_{i_j}(t) < \theta_i \\ \sigma_i(t), & \sum_{j=1}^{k_i} a_{i,j}\sigma_{i_j}(t) = \theta_i, \end{cases} \quad (2)$$

where $\{\sigma_{i_1}, \ldots, \sigma_{i_{k_i}}\}$ are the k_i regulators of σ_i. The interaction strength (or weight) $a_{i,j}$ takes a positive (or negative) value if σ_{i_j} is an activator (or a repressor) of σ_i, respectively.[3] The activation threshold θ_i of σ_i indicates the minimum value of the sum required for the activation of the node to take place. The dynamic rule given in Eq. (2) is the same as the one used in Refs. [15,16]. However, in that work the authors considered the simple case in which $a_{i,j} = 1$ for activators, $a_{i,j} = -1$ for repressors, and $\theta_i = 0$ for almost all the nodes except by a few ones. Additionally, "self-degradation" was introduced to some of the nodes just by making $a_{i,i} = -1$.

The number k_i of regulators for each node σ_i is drawn from a probability distribution $P_{in}(k)$, and then these regulators $\{\sigma_{i_1}, \ldots, \sigma_{i_{k_i}}\}$ are randomly chosen from anywhere in the system. Each regulatory interaction strength $a_{i,j}$ is set activatory with probability p and inhibitory with probability $1-p$. All activatory interactions have a value $a_{i,j} = a_G$, whereas the inhibitory interactions have

---
[3] Ofcourse, $a_{i,j} = 0$ if there is no interaction between σ_{i_j} and σ_i.

a value $a_{i,j} = -a_R$, where a_G and a_R are positive integers. The ratio a_G/a_R measures the relative importance of activation over repression. Thus, if a_G/a_R is small, then inhibitory interactions are dominant, whereas if a_G/a_R is large, then activation dominates over repression. Finally, we set a fixed value of the activation threshold $\theta_i = \theta$ for all nodes. We consider three cases corresponding to three different threshold values: $\theta = 0.5$, $\theta = 0$ and $\theta = -0.5$. The rationale for this choice is two fold. First, these values suffice to illustrate the effects of integer and non-integer thresholds on the dynamics. And second, because these are the values that have been used in models of real genetic networks, obtaining good agreement with experimental observations [15,16].

On thing that should be noted from Eq. (2) is the effect that the value of the threshold θ_i has on the dynamics. If we consider only integer values for the interaction strenghts $a_{i,j}$, then the equality in Eq. (2) can be attained only if θ_i is also an integer. In such a case, the last row on the right-hand side of Eq. (2) implies that every node regulates itself. In other words, given the interaction strengths $a_{i,j}$ and the thresholds θ_i, the right-hand side of Eq. (2) can be written as a Boolean function F_i only if we assume that σ_i belongs to its own set of regulators. Because of this, we have explicitely written $\sigma_i(t)$ as one of the arguments of the regulatory function F_i. This self-regulation does not necessarily happen in Kauffman Boolean networks, and it can make a big difference with regard to the dynamical behavior. As we will see below, the fact that integer threshold values allow the node σ_i to simply stay in their previous state and essentially freeze plays a mayor part in the dynamical behaviour of the network and in its use for biological modeling.

Note also from Eq. (2) that all the information necessary for the network dynamics is contained in a N-dimensional vector $\boldsymbol{\theta} = (\theta_1, \theta_2, \ldots, \theta_N)$ whose components are the thresholds, and a $N \times N$ matrix $\mathbf{A}$. This matrix is such that $[\mathbf{A}]_{i,j} = a_{i,j}$ if σ_{i_j} is a regulator of σ_i, and $[\mathbf{A}]_{i,j} = 0$ otherwise. This is very different from what happens in RBN's, where to store all the information necessary for the network dynamics we need a $N \times N$ matrix containing the topology of the network, and a Boolean function F_i for each node σ_i. Each of these functions has 2^{k_i} entrances, one for each configuration of its k_i inputs. As mentioned before, for a given set of thresholds and interaction strengths, we can also create a Boolean function corresponding to the rule given in Eq. (2). However, this limits the set of possible Boolean functions that can be obtained.

2.2 The Derrida Map: Deviations from the Kauffman Behaviour

One of the most useful ways to study the general dynamics of Boolean networks has been in terms of the propagation of perturbations (also called damage spreading) throughout the network. To this end, let us denote as Σ_t the dynamical configuration of the network at time t, that is, $\Sigma_t = \{\sigma_1(t), \sigma_2(t), \ldots, \sigma_N(t)\}$. Let Σ_0 and $\widetilde{\Sigma}_0$ be two slightly different intitial configurations, namely, $\widetilde{\Sigma}_0$ is almost identical to Σ_0 except by a few nodes which have reversed their values (this is the initial perturbation or the initial damage). Under the dynamics given in

Eq. (2), each of these initial configurations will generate a trajectory throughout time:

$$\Sigma_0 \to \Sigma_1 \to \cdots \to \Sigma_t \to \cdots$$
$$\widetilde{\Sigma}_0 \to \widetilde{\Sigma}_1 \to \cdots \to \widetilde{\Sigma}_t \to \cdots$$

These two trajectories may eventually converge (the initial perturbation disappears), diverge (the initial perturbation amplifies), or remain "parallel" (the initial perturbation neither grows nor disappears). These three different behaviors determine the dynamical regime in which the network operates: In the ordered regime the two trajectories typically converge after a transient time. In the chaotic regime the system becomes very sensitive to small changes in the initial condition and the two trajectories diverge from each other. The intermediate case where, on average, perturbations retain their same size corresponds to the so called critical regime. The critical regime has been extensively studied and appears to be characteristic property of genetic networks [30,31,32,33,34].

We quantify the propagation of perturbations in the network in terms of the time evolution of normalized Hamming distance $h(t)$, which is defined as

$$h(t) = d\left(\Sigma_t, \widetilde{\Sigma}_t\right) = \frac{1}{N}\sum_{i=1}^{N}|\sigma_i(t) - \widetilde{\sigma}_i(t)|. \tag{3}$$

The assymptotic value $h_\infty = \lim_{t\to\infty} h(t)$ is the final size of the avalanche of perturbations and acts as the order parameter of the system: In the ordered regime $h_\infty = 0$, while in the chaotic regime $h_\infty > 0$. In the critical regime $\lim_{t\to\infty} h(t) = 0$ only marginally, which means that it can take a long time for a small perturbation to disappear.

For a given network realization, h_∞ can be computed numerically in two different ways. The first way is a direct implementation of the definition. We start out the dynamics from two different initial conditions Σ_0 and $\widetilde{\Sigma}_0$, and let the system evolve for a long time t_r. Then, h_∞ is the Hamming distance $h(t_r)$ between the two final configurations Σ_{t_r} and $\widetilde{\Sigma}_{t_r}$, averaged over many pairs of initial conditions. We will denote the value of the order parameter obtained by this method as $h_\infty^{(1)}$.

The second way to compute h_∞ is by means of the so-called Derrida map $M(h)$ [35], which relates the size of a perturbation avalanche between two consecutive time steps, that is, $h(t+1) = M(h(t))$. Starting from two different initial configurations whose Hamming distance is h_0, successive iterations of this map eventually converge to h_∞. Thus, h_∞ is the stable fixed point of the Derrida map: $h_\infty = M(h_\infty)$. For RBN's, mean-field theory computations show that $M(h)$ is a continuous convex monotically increasing function with the properties $M(0) = 0$ and $M(1) < 1$. For threshold networks this mapping is still continuous and satisfies $M(0) = 0$ and $M(1) < 1$, but it is not clear whether or not it is a monotonically increasing function. Nonetheless, for the set of parameters we use in this work $M(h)$ seems to satisfy all the properties predicted by the mean-field theory. The fulfillment of these properties is important because this guarantees the existence of one and only one stable fixed point. In this case, the dynamical regime in which the network operates is determined by the slope at the origin of $M(h)$, called the *average network sensitivity S*:

$$S = \left.\frac{dM(h)}{dh}\right|_{h=0}. \qquad (4)$$

If $S < 1$ then $h_\infty = 0$ and the system is in the ordered phase, whereas if $S > 1$ then $h_\infty > 0$ and the system is in the chaotic regime. The critical regime is attained for $S = 1$, which is the point at which the phase transition between the ordered and chaotic regimes occur [35,36].

To compute $M(h)$ numerically for a given network realization, we start from two different configurations Σ_0 and $\widetilde{\Sigma}_0$ separated by a Hamming distance h_0. Next, we evolve these two initial configurations just one time step and compute the Hamming distance h_1 between the resulting configurations Σ_1 and $\widetilde{\Sigma}_1$. The value $M[h_0]$ of the Derrida map at h_0 is then obtained by averaging h_1 over many pairs of initial conditions whose Hamming distance is h_0. By doing this for all values of $h_0 \in (0,1)$ we can construct the full curve $M(h)$ and compute its fixed point h_∞. We will denote the value of the order parameter obtained by this method as $h_\infty^{(2)}$.

For general RBN's it has been shown that $h_\infty^{(1)}$ and $h_\infty^{(2)}$ are very close to each other. Actually, in the thermodynamic limit $N \to \infty$ they are the same [35]. The reason for this is that in RBN's the temporal correlations between two consecutive configurations Σ_t and Σ_{t+1} are inversely proportional to the number of nodes N. Therefore, for large networks with completely random Boolean functions the mean-field conditions are satisfied and the temporal evolution is essentially dependent on the previous time step only. However, when temporal correlations extend over several time steps, the Derrida map does not accurately predict the value of the order parameter. In such cases $h_\infty^{(1)}$ and $h_\infty^{(2)}$ can differ by a large ammount. This non-ergodic behavior in the network dynamics has been observed in Boolean networks in which only a small subset of the class of all Boolean functions are used [37], such as canalyzing functions [29] and threshold functions with equal values and proportions of activation/repression strengths [27]. For the general case of RTN's we also observe a large deviation from the ergodic behavior assumed by the mean-field computation.

In Fig. 1 we plot $h_\infty^{(1)}$ (diamonds) and $h_\infty^{(2)}$ (circles) as functions of the network connectivity K, for RBN's (Fig. 1a) and RTN's (Fig. 1a,c). We also plot the quantity h_∞^* predicted analytically using the annealed approximation presented in Sec. 3, which is a generalization of the one reported in Ref. [27].[4] Note from Fig. 1a that for RBN's the three values of h_∞ are identical within numerical accuracy, which reflects the ergodicity of the system in this case. However, for RTN's such ergodicity dissappears, as it is apparent from the fact that $h_\infty^{(1)}$ is

---

[4] This computation incorporates in an approximate way the temporal correlations between succesive network states using the final number of active and inactive nodes.

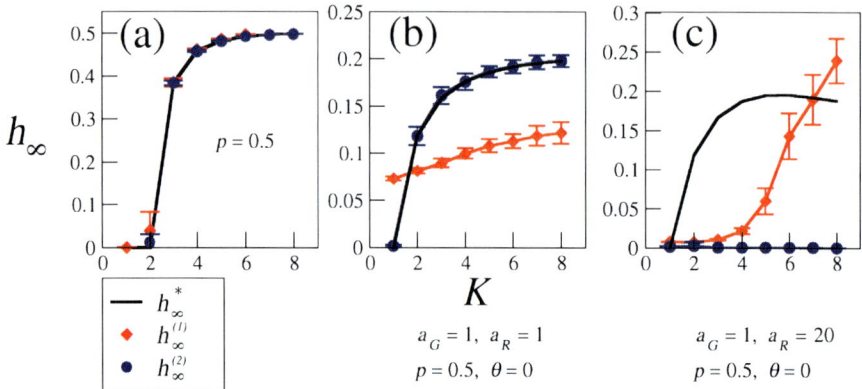

Fig. 1. The nonergodicity of the system is illustrated by plotting the order parameter $h_\infty^{(1)}$ computed directly from the definition (red diamonds), and the order parameter $h_\infty^{(2)}$ computed as the fixed point of the Derrida map (blue circles). The analytic prediction h_∞^* from the annealed computation presented in see Sec. 3 is also plotted (solid line). Three different ensembles of networks are used: (a) Standard Kauffman networks (RBN's). In this case $h_\infty^{(1)} = h_\infty^{(2)} = h_\infty^*$, which shows that RBN's are ergodic. (b) Random Threshhold Networks (RTN's) with $p = 0.5$, $a_G = a_R = 1$ and $\theta = 0$. Note in this case that $h_\infty^{(1)} \neq h_\infty^{(2)}$ (although $h_\infty^{(2)} = h_\infty^*$), which reflects the nonergodicity of the dynamics. Finally, (c) corresponds to RTN's with $p = 0.5$, $a_G = 1$, $a_R = 20$ and $\theta = 0$. In this last case, not only is $h_\infty^{(1)} \neq h_\infty^{(2)}$, but also the analytic prediction h_∞^* fails completely. In all cases, each point is the average over 100 network realizations, each having $N = 1000$ nodes. For each of these networks we used 10000 pairs of random initial conditions.

quite different from $h_\infty^{(2)}$. Fig. 1b corresponds to the case in which $\theta = 0$ for all nodes and the weights take the values $a_{i,j} = \pm 1$, chosen with equal probability. This strong deviation is surprising, especially since it happens for the simplest case similarly to the one used in Refs. [15,16] for the modelling of the yeast cell-cycle networks. Furthermore, departure from ergodicity is even worse for unequal weights, as shown in Fig. 1c, where the negative weights were chosen to be ten times stronger than the positive weights, i.e. $a_R = 10$ and $a_G = 1$. In this case $h_\infty^{(1)}$ does not only deviates from $h_\infty^{(2)}$, but also the analytical prediction h_∞^* completely fails to reproduce $h_\infty^{(1)}$. The above results indicate that some care must be taken when choosing the parameters in RTN's if these networks are to be used for biological modelling. We explore this issue furtherly in the next sections.

3 The Phase Diagram

The Derrida map $M(h)$ can be computed analytically within the context of the so-called *annealed approximation*, first introduced by Derrida and Pomeau [28].

This mean-field technique assumes statistical independence between the nodes and neglects the temporal correlations developed throughout time between succesive states of the network. The annealed approximation has been successfully used in RBN's to obtain analytically where the phase transition occurs for different topologies and network parameters [36,8]. However, the mean-field assumptions fail dramatically for RTN's as it is illustrated in Fig. 1. In an attempt to improve the annealed approximation, one has to incorporate into the analysis the temporal correlations between succesive network states [29,37]. This has been done for particular values of the parameters [27]. Here we present a generalization of this computation valid for different activator/repressor strengths, ratios and thresholds.

We start the computation of $M(h)$ by introducing the quantity $I^{(k_d)}$, known as the *influence of k_d variables*. Let us consider an arbitrary network in the ensemble of RTN's, and pick out a node σ_i with k_i inputs. Let Σ_t and $\widetilde{\Sigma}_t$ be two network configurations in which k_d of the inputs of σ_i (with $k_d \leq k_i$) have been damaged, namely, these k_d inputs have opposite values in these two configurations.[5] $I^{(k_d)}$ is defined as the probability that this initial damage of k_d inputs propagates one time step, which means that σ_i will have different values in Σ_{t+1} and $\widetilde{\Sigma}_{t+1}$. These influences do not only depend on the ensemble of Boolean functions used, but have also been shown to depend heavily on the bias in the expected probability with which the system visits the different states of its configuration space. In previous work, this bias has been expressed in terms of the fraction $b(t)$ of active nodes in the system [27,38,37].

By using the annealed approximation assumptions, in Appendix A we show that the temporal evolution of $b(t)$ is given by

$$b(t+1) = B\left(b(t)\right) = p_+ \left(b(t)\right) + b(t) \cdot p_0 \left(b(t)\right), \tag{5a}$$

where p_+ and p_0 are the probabilities that, for a given node, the sum of its inputs is larger than or equal to the threshold θ, respectively. These probabilities can be written as (see Appendix B)

$$p_+ \left(b(t)\right) = \sum_{k_i=1}^{\infty} P_{in}(k_i) \sum_{i=0}^{k_i} \binom{k_i}{i}(1-b)^i b^{k_i-i} \sum_{l=l_i}^{k_i-i} \binom{k_i-i}{l} p^l q^{k_i-i-l}, \tag{5b}$$

$$\text{where } l_i = \left\lceil \frac{(k_i-i)a_G + \theta}{(a_G + a_R)} \right\rceil + 1$$

$$p_0 \left(b(t)\right) = \sum_{k_i=1}^{\infty} P_{in}(k_i) \sum_{i=0}^{k_i} \binom{k_i}{i}(1-b)^i b^{k_i-i} \sum_{l=0}^{k_i-i} \binom{k_i-i}{l} p^l q^{k_i-i-l}$$

$$\times \delta_{a_G l, a_R(k_i-i-l)+\theta}. \tag{5c}$$

---

[5] Given that statistical equivalence is assumed, then σ_i will be representative of the entire network. Therefore, only the state of the inputs of σ_i is important, regardless of the states of all the other nodes.

Since we are interested in the asymptotic value h_∞ of the order parameter, it is necessary to compute the final number of active elements $b_\infty = \lim_{t\to\infty} b(t)$. This is just the stable fixed point of the map given in Eq. (5a), namely, $b_\infty = B(b_\infty)$. From the set of Eqs. (5), the value of b_∞ is computed numerically for each particular realization of parameters. Once the value of b_∞ is known, it is used to compute the influence $I^{(k_d)}$. In Appendix C we show that $I^{(k_d)}$ is then given by

$$I^{(k_d)} = \sum_{i=0}^{k_i} \binom{k_i}{i} p^i q^{k_i-i} \sum_{l=0}^{i} \sum_{m=0}^{k_i-i} \binom{i}{l}\binom{k_i-i}{m} b_\infty^{l+m} (1-b_\infty)^{k_i-l-m}$$
$$\times \mathcal{I}(k_i, k_d, i, l, m), \qquad (6)$$

where $\mathcal{I}(k_i, k_d, i, l, m)$ is defined as

$$\mathcal{I} = \sum_{u=u_0}^{u_f} \sum_{v=v_0}^{v_f} \sum_{w=w_0}^{w_f} \frac{\binom{l}{u}\binom{m}{v}\binom{i-l}{w}\binom{k_i-i-m}{k_d-u-v-w}}{\binom{k_i}{k_d}}$$
$$\times \{ H(a_G(l-u+w) - a_R(m-v+z) - \theta) \cdot [(1-b_\infty)\delta_{a_G l - a_R m, \theta}$$
$$+ H(a_R m + \theta - a_G l)] + \delta_{a_G(l-u+w), a_R(m-v+z)+\theta} \cdot [h(t)\delta_{a_G l - a_R m, \theta}$$
$$+ b_\infty H(a_R m + \theta - a_G l) + (1-b_\infty) H(a_G l - a_R m - \theta)]$$
$$+ H(a_R(m-v+z) + \theta - a_G(l-u+w)) \cdot [b_\infty \delta_{a_G l - a_R m, \theta}$$
$$+ H(a_G l - a_R m - \theta)]\}. \qquad (7)$$

where the summation is done between the limits of a multivariate hypergeometric distribution,[6] and $H(x)$ is the Heaviside step function with $H(0) = 0$.

Note that the influence $I^{(k_d)}$ already contains information about the temporal correlations through the value of b_∞. However, the above expressions are not exact because b_∞ is computed from Eqs. 5, which were formulated using the mean-field assumptions. In spite of this approximation, it is an improvement over the original annealed approximation which completely neglects the temporal correlations. Once the value of $I^{(k_d)}$ if obtained from the above equations, it is used to obtain the Derrida map, which determines the temporal evolution of the Hamming distance, as [29]

$$h(t+1) = M(h(t)) = \sum_{k_i=1}^{\infty} P_{in}(k_i) \sum_{k_d=0}^{k_i} I^{(k_d)} \binom{k_i}{k_d} [h(t)]^{k_d} [1-h(t)]^{k_i-k_d}. \qquad (8)$$

This equation tells us that the size of a perturbation avalanche after one time step depends on the probability to find k_d damaged input nodes between two configurations Σ_t and $\widetilde{\Sigma}_t$, and on the probability $I^{(k_d)}$ that this damage spreads to the configurations Σ_{t+1} and $\widetilde{\Sigma}_{t+1}$ at the next time step.

---

[6] Specifically we have that $u_0 = \max(0, k_d + l - k_i)$, $u_f = \min(l, k_d)$; $v_0 = \max(0, k_d - u - (k_i - l - m))$, $v_f = \min(l, k_d - u)$; $w_0 = \max(0, k_d - u - v - (k_i - l - m - i + l))$, $w_f = \min(l, k_d - u - v)$.

3.1 Sensitivity and Influence of 0 Variables

Once b_∞ is obtained from Eq. (5a), it is possible to calculate the phase diagrams for the different parameters involved in a network realization using Eqs. (6), (8) and (4). There is however one last point that needs to be considered before the computation of the sensitivity, which is that the average influence of 0 variables $I^{(0)}$ is not necessarily null.

By definition, $I^{(0)}$ is the probability that, for a given node σ_n, a damage on none of its input elements spreads to the next time step. This means that σ_n will have different values in the configurations Σ_{t+1} and $\widetilde{\Sigma}_{t+1}$ even when all of its inputs had the same values in the previous configurations Σ_t and $\widetilde{\Sigma}_t$. In Kauffman RBN's this cannot happen, because the equality of the inputs of σ_n in the two configurations Σ_t and $\widetilde{\Sigma}_t$ guarantees that σ_n will have the same value in the next configurations Σ_{t+1} and $\widetilde{\Sigma}_{t+1}$, and therefore, in this case $I^{(0)} = 0$.[7] However, for RTN's, the last line in Eq. (2) makes it possible for σ_n to be different in Σ_{t+1} and $\widetilde{\Sigma}_{t+1}$ even when all of its inputs were the same in the previous configurations Σ_t and $\widetilde{\Sigma}_t$. This happens when $\sum_j a_{n,j}\sigma_{n_j}(t) = \theta$ and σ_n had different values in the configurations Σ_t and $\widetilde{\Sigma}_t$. In such a case σ_n will remain different in the next configurations Σ_{t+1} and $\widetilde{\Sigma}_{t+1}$. This can be considered as a damage spread for zero input variables, and consequently $I^{(0)} \neq 0$. Note that this happens only when the equality in the last line in Eq. (2) is satisfied, which in turn occurs only for integer values ot θ.

Taking the above considerations into account, it is possible to write $I^{(0)}$ as

$$I^{(0)} = p_0(b_\infty)\, h(t) \quad \text{with } \theta \in \mathbb{Z}, \tag{9}$$

Using the previous equation, one is finally able to get the sensitivity of the network, defined in Eq. (4), as

$$S = p_0(b_\infty) + \sum_{k_i=1}^{\infty} P_{in}(k_i)\, k_i\, I^{(1)}. \tag{10}$$

The derivation of the last two expressions is presented in Appendix D.

Note that $I^{(1)}$ also depends on k_i, and that both $I^{(1)}$ and $p_0(b_\infty)$ depend on the parameters of the network realization p, a_G, a_R and θ. Interestingly, the sensitivity S, and thus, the dynamical phase in which the system operates, only depends on the lower influences $I^{(0)}$ and $I^{(1)}$. This means that the effect of small changes in the two configurations Σ_t and $\widetilde{\Sigma}_t$ are the ones that determine the newtork dynamical regime. However, Σ_t and $\widetilde{\Sigma}_t$ cannot be arbitrary, since they must have a fraction of active elements close to the final one, b_∞. This restriction has profound effects on the initial apparent dynamical behavior of the network as compared to what actually happens at the end of the dynamics, in the sense that two trajectories that initially appear to converge may end up diverging, and vice versa. We will discuss this problem in Sec. 4.

---

[7] This is why in Ref. [29] the summation over k_d excludes $k_d = 0$, whereas in our Eq. (8) the sum starts from $k_d = 0$.

3.2 The Phase Diagram for the Homogeneous Random Topology

Eq. (10) determines the structure of the phase diagram as a function of the network topology (contained in $P_{in}(k)$), and the other parameters of the system. Here we consider the homogeneous random topology $P_{in}(k) = \delta_{K,k}$ in which each node has exactly K regulators randomly chosen from anywhere in the system. In such a case, Eq. (10) establishes a relationship between the sensitivity S and the value of the parameters K, p, a_G, a_R and θ. The ordered phase occurs in those regions of the parameter space in which $S < 1$, whereas the chaotic phase occurs whenever $S > 1$. The critical region is the one for which $S = 1$. As the parameter space is 5-dimensional, an exhaustive exploration is neither illustrative nor computationally feasible. Instead, we present the phase diagram K vs. p for the following cases, which are representative of the general behavior observed across the entire parameter space:

- **Case 1:** Activating and inhibiting interactions are of the same magnitude ($a_G = 1$, $a_R = 1$);
- **Case 2:** Inhibiting interactions are stronger than activating ones ($a_G = 1$, $a_R = 2$);
- **Case 3:** Activating interactions are stronger than inhibiting ones ($a_G = 2$, $a_R = 1$);
- **Case 4:** Inhibiting interactions are always dominant ($a_G = 1$, $a_R = 20$);
- **Case 5:** Activating interactions are always dominant ($a_G = 20$, $a_R = 1$);

Additionally, for each of the five cases listed above we used the threshold values $\theta = 0.5$, $\theta = 0$ and $\theta = -0.5$.

The resulting phase diagrams are shown in Fig. 2. It is immediately apparent from this figure the asymmetric structure of the phase diagram with respect to the activator fraction (measured by p) and strength (measure by the quotient a_G/a_R). In general, it appears that activators strongly push the network into the frozen phase (in blue), while repressors move it towards the chaotic phase (in red), but less drastically. This can be seen in the extreme cases of dominant activators ($a_G = 20$, $a_R = 1$) where the chaotic region almost dissappears, while for the opposite case of dominant repression ($a_G = 1$, $a_R = -20$) the frozen region is considerably smaller than the chaotic one. Another important point to note is the different behavior of the critical line for the three threshold values of interes: For $\theta = 0.5$ there are two critical values of p for each value of K, whereas for $\theta = -0.5$ and $\theta = 0$ there is only one. In this sense, of all the cases shown in Fig. 2, the phase diagrams for $\theta = 0.5$ are the ones closer to the phase diagram obtained for RBN's [8].

Finally, it is important to mention that we obtain the same results reported in Ref. [27] for the special case $p = 0.5$, $a_G = 1$ and $a_R = 1$, but only for the threshold values $\theta = 0.5$ and $\theta = -0.5$. However, for $\theta = 0$ we obtain a completely different behavior as the one reported in Ref. [27]. Indeed, we find that the phase transition occurs at $K = 1$, whereas the authors in Ref. [27]

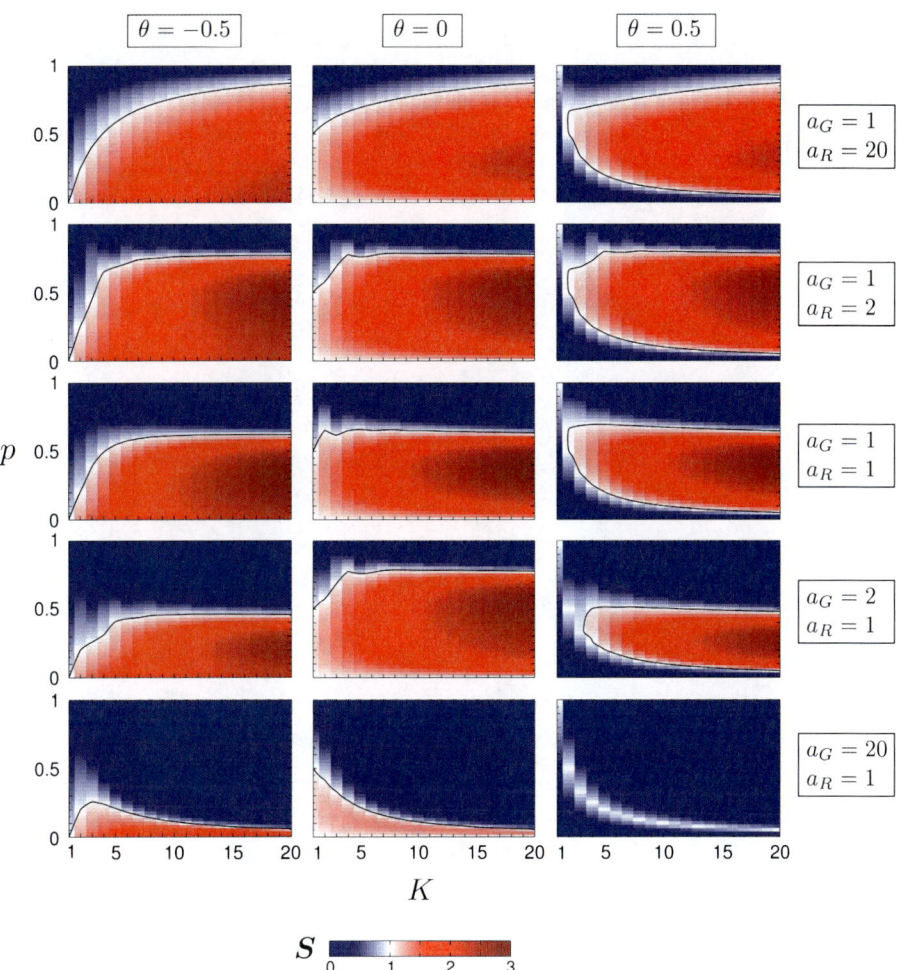

Fig. 2. Phase diagram p vs. K for threshold networks with different parameters, obtained by numerically solivng Eq. (10). The color code represents the value of the sensitivity S. The ordered phase ($S < 1$) is represented in blue while the chaotic phase ($S > 1$) is represented in red. Zones near the critical region ($S = 1$) are white, while the critical region itself is represented by the black line. Note the asymmetry of the phase diagrams, especially for $\theta = 0$ and $\theta = -0.5$.

report that the phase transition ocurs between $K = 12$ and $K = 13$. This discrepancy is due to not properly taking into account the self-regulation conveyed in the last line of Eq. (2), which happens only for integer threshold values (see Sec. 2.1). In the next section we present numerical results that support our analytic approach for integer threshold values.

4 Numerical Experiments

To test the validity of the expressions obtained in Sec. 3 we performed numerical simulations of the network dynamics using ensembles of 100 RTN's, with $N = 1000$ nodes each. We used networks with homogeneous random topologies and varied K from $K = 1$ to $K = 8$. The other parameters p, a_G, a_R and θ were chosen to represent the different behaviors depicted in the phase diagrams shown in Fig. 2.

4.1 The Uncorrelated Sensitivity S_0 and the Final Avalanche Size h_∞

To compare the analytical expressions with the results of the numerical simulation we need to compute two parameters: The *uncorrelated network sensitivity* S_0 and the *final avalanche size* $h_\infty^{(1)}$. In Sec. 2.2 we describe how to compute $h_\infty^{(1)}$ for a given network realization. To compute S_0 let us consider two initial configurations Σ_0 and $\widetilde{\Sigma}_0$ that differ only in one element, namely, whose Hamming distance is $1/N$:

$$d\left(\Sigma_0, \widetilde{\Sigma}_0\right) = \frac{1}{N}. \tag{11}$$

Then, S_0/N is the average Hamming distance of the two configurations at the next time step:

$$S_0 = N \left\langle d\left(\Sigma_1, \widetilde{\Sigma}_1\right) \right\rangle,$$

where the average $\langle \cdot \rangle$ is taken over all possible pairs of initial conditions satisfying Eq. (11). In other words, S_0 is the average size of the perturbation avalanche after one time step, given an initial perturbation of only one node. Note that S_0 is the slope at the origin of the Derrida map without taking into account the correlations developed throughout time between network states. This is the reason why we call S_0 the "uncorrelated" sensitivity. Therefore, for large N, S_0 should be the sensitivity of the network given by Eq. (10) with $b_\infty = b_0 = 0.5$, which assumes complete independence between the network states.[8]

Since the uncorrelated sensitivity S_0 has no dependence on the correlations developed in time, we expect the analytic results derived in Sec. 3 to accurately reproduce the behavior of S_0. However, we do not expect this analytic computation to describe as acurately the value of $h_\infty^{(1)}$, because in this computation the temporal correlations were approximately taken into account only through the value of b_∞, whereas the actual value of $h_\infty^{(1)}$ depends on the precise way in which the network evolves in time. This is illustrated in Figs. 3 and 4.

It can be seen in Fig. 3 that the uncorrelated sensitivity S_0 computed numerically (symbols) shows a remarkable agreement with the theoretical prediction

---

[8] Since the initial configuration Σ_0 in Eq. (11) is chosen randomly from all possible configurations, the sequence of 0's and 1's in Σ_0 can be thought of as N independent Bernoulli trials with probability $1/2$, which gives $b_0 = 0.5$ for the expected fraction of 1's.

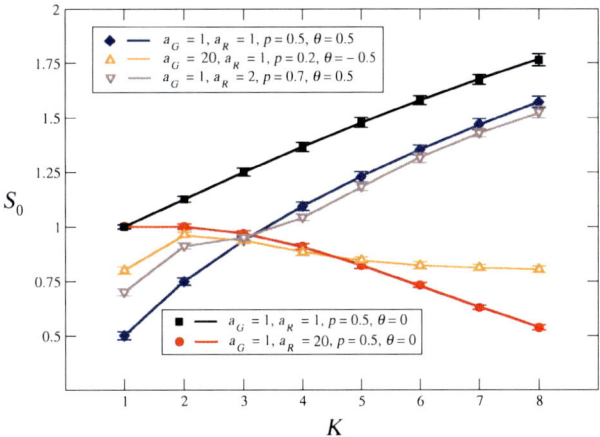

Fig. 3. Uncorrelated sensitivity S_0 for RTN's with $N = 1000$ and different values of the parameters p, a_G, a_R and θ. The symbols are numerical data computed using ensembles of 100 networks. For each of these networks, S_0 was averaged over 10000 pairs of initial conditions differing in just one node. The error bars represent the standard deviation. The solid lines correspond to the theoretical result gien in Eq. (10) with $b = 0.5$. Note the excellent agreement between the theoretical prediction and the numerical data for all the different parameters used.

(solid line) for the different combinations of parameters used. It is interesting to note the variety of behaviours exhibitted by S_0 in RTN's, which is in marked constrast with the linear behavior observed in standard Kauffman Nets. Indeed, for RBN's $S_0 = 2p(1-p)K$, whereas for Kauffman networks with canalyzing functions $S_0 = 1/2 + (K-1)/4$ [29]. In contrast, Fig. 3 shows that for RTN's the dependance of S_0 on K is nonlinear and can even change inflexion or decrease with increasing K. This general nonlinear behaviour occurs even for the simple cases $p = 0.5$, $a_G = a_R = 1$, $\theta = 0$, and $\theta = \pm 0.5$, where we have (see Appendix E for a derivation)

$$S_0 = \begin{cases} \dfrac{K+1}{2^{2K}}\dbinom{2K}{K} & \text{for } \theta = 0 \\[2ex] \dfrac{K}{2^{2K}}\dbinom{2K}{K} & \text{for } \theta = \pm 0.5 \end{cases} \qquad (12)$$

in which $S_0 \sim \sqrt{K}$ for large K. The above result allows the network to ramain close to the critical phase for a wider range of values of K than standard RBN's. This might be important given that there is evidence showing that real genetic networks, in which the gene input connectivity varies considerably from one gene to another, work near the critical phase $S_0 = 1$ [34].

With regard to the final size of the perturbation avalanche, Fig. 4 shows the value $h_\infty^{(1)}$ computed numerically (line with symbols), and the value h_∞^* predicted

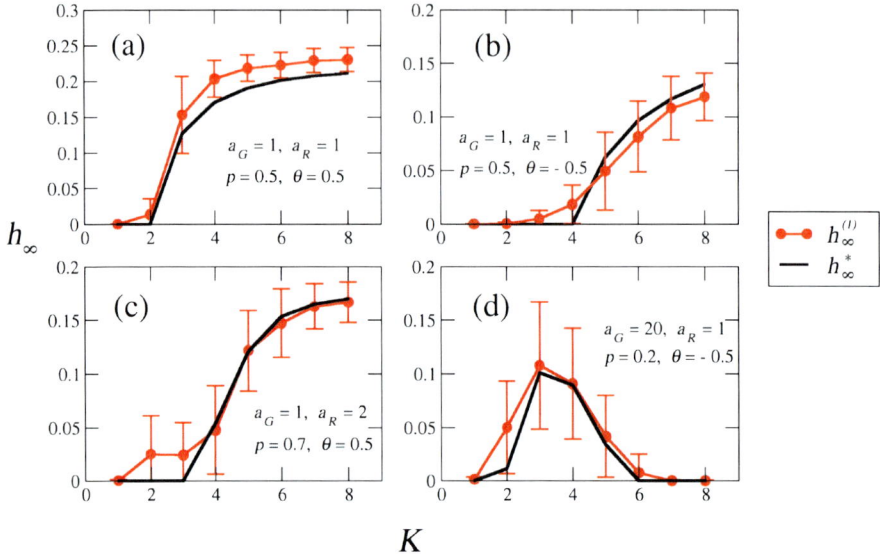

Fig. 4. Final size of the perturbation avalanche computed numerically ($h_\infty^{(1)}$, red circles), and analytically as the fixed point of Eq. (8) (h_∞^*, solid black line). The numerical data were computed for an ensemble of 100 RTN's with $N = 1000$ and different values of the parameters p, a_G, a_R and θ. The important point in this figure is the use of noninteger threshold values: $\theta = \pm 0.5$. For each network realization, we averaged $h_\infty^{(1)}$ over 10000 pairs of random initial conditions with Hamming distance $h_0 = 0.1$. Note the large standard deviations in the numerical dada (error bars). Despite this enormous variability in each network realization, the average numerical data qualitatively follow well the theoretical prediction.

by the analytic computation of Sec. 3 (solid line). Although $h_\infty^{(1)}$ and h_∞^* are qualitatively very similar to each other for the cases $\theta = \pm 0.5$ depicted in Fig. 4, their quantitative correspondence is not as precise as it was for S_0. As it was mentioned before, this lack of precision was expected due to the approximation in the computation of the temporal correlations. However, for integer threshold values $h_\infty^{(1)}$ and h_∞^* do not necessarily agree even qualitatively, as it is illustrated in Fig. 1 for the special case $\theta = 0$. We discuss the origing of this discrepancy further below. In the mean time, it is important to emphasize the reason why the Derrida map is not always useful to discriminate the dynamical regime in which the network operates.

Fig. 6 shows the temporal evolution of the average Hamming distance $h(t)$ between two trajectories that started from two slightly different initial conditions Σ_0 and $\widetilde{\Sigma}_0$. (The average is taken over many pairs of initial conditions.) In all the cases shown in this figure, $h(t)$ decreases in the first time steps. Therefore, according to the Derrida map, which takes into account only the first time step, these networks should be in the ordered regime. However, after that initial decrease, the Hamming distance $h(t)$ increases again reaching a value considerably larger than the initial Hamming distance $h(0)$. Thus, in the long term the initial perturbation is amplified, which means that the dynamical regime in which these networks operate turns out to be chaotic. It should be noted that the behavior reported in Fig. 6 was obtained for networks with biologically reasonable values of the parameters: $a_G = a_R = 1$, $\theta = 0.5$, $p = 0.5$ and $K = 3$.

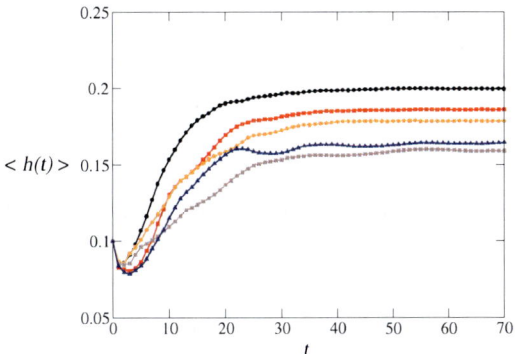

Fig. 5. Temporal evolution of the Hamming distance for 5 different random threshold network realizations with $N = 1000$, $a_G = a_R = 1$, $\theta = 0.5$, $p = 0.5$ and $K = 3$. Each curve is the average over 10000 randomly chosen pairs of initial conditions separated by a Hamming distance $h_0 = 0.1$. Note that initially the Hamming distance decreases. Using only the Derrida map, which takes into account only the first time step, one would conclude that the networks operate in the ordered regime. However, the correlations developed in time due to the network structure and the number of active nodes make the Hamming distance rise again and approach a nonzero value, which is characteristic of the chaotic regime.

4.2 The $\theta = 0$ Case

We now address the anomalous case $\theta = 0$. As we have seen in the previous section, in this case the annealed approximation gives very accurate results for the initial sensitivity S_0 but very poor results for the final avalanche size $h_\infty^{(1)}$. As discussed in Sec. 2, integer thresholds allow the possibility for some nodes to become frozen whenever their input sum in Eq. (2) equals the threshold. These frozen nodes generate explicit correlations in time, which in turn produce a strong dependance on the history of the dynamics, and thus, on the initial conditions. This is illustrated in Fig. 6 where the final avalance size $h_\infty^{(1)}$ is plotted against the initial perturbation size $h_0 = h(0)$ for networks with $p = 0.5$, $a_G = a_R = 1$ and $\theta = 0$. It is apparent from this figure that for $K \leq 5$ the value of $h_\infty^{(1)}$ strongly depends on h_0, and this dependece becomes less strong as K increases. This is because for large values of K it is harder for the input sum to equal the threshold.

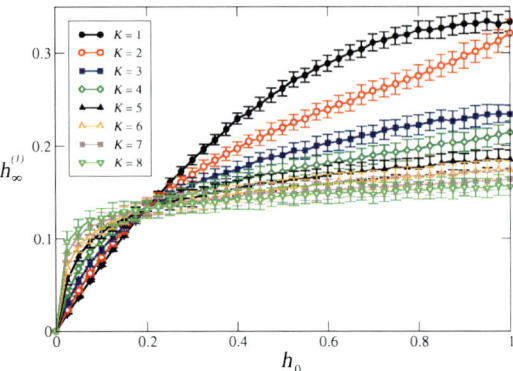

Fig. 6. Final size $h_\infty^{(1)}$ of the perturbation avalanche as a function of the initial perturbation size h_0. Each point is the average over 100 random threshold network realizations with $N = 1000$, $a_G = a_R = 1$, $p = 0.5$, $\theta = 0$, and 1000 pairs of random initial conditions for every h_0 in each of these networks. Note the strong dependence of $h_\infty^{(1)}$ on h_0, especially for small values of K where the temporal correlations are stronger. For large values of K these correlations become weaker and consequently $h_\infty^{(1)}$ becomes almost independent of h_0.

Another problematic consequence of using integer threshold values is the existence of an enormous number of punctual attractors, many of which differ only in the value of just one node. This anomaly has been noted before in Ref. [27]. It also occurs in the cell cycle models of *S. cerevisiae* [15] and *S. pombe* [16], where many punctual attractors with small basins of attraction were found.[9] Fig. 7 shows the average number of attractors as a function of the network connectivity K for $\theta = 0$ (circles), $\theta = 0.5$ (squares) and $\theta = -0.5$ (triangles). In all cases we used $p = 0.5$ and $a_G = a_R = 1$. Fig. 7a corresponds to large networks with $N = 1000$. In this case the number of possible configurations is astronomically huge ($\Omega = 2^{1000}$). Therefore, an under sampling of the state space has to be done, in which case only some attractores will be found. We sampled 5×10^4 configurations. Surprisingly, for $\theta = 0$ *almost every sampled initial configuration ended up in a different attractor*. The same happens for smaller networks with $N = 100$, as it is shown in Fig. 7b. However, the same undersampling performed in networks with non-integer threshold values ($\theta = \pm 0.5$) reveals a number of atractors which is several orders of magnitude smaller than the one obtained for $\theta = 0$. Finally, Fig. 7c shows similar results but for small networks with $N = 15$, for which the entire state space can be probed ($\Omega = 2^{15} = 32768$). Note that in this case, the average number of attractors decreases with K for $\theta = 0$. This behavior is marked contrast with the one observed for non-integer threshold values (and for RBN's), where the average number of attractors grows with the network connectivity K.

---

[9] However, in these cell cycle models the authors deemed the attractors with small basins of attractions as biologically irrelevant and neglected them.

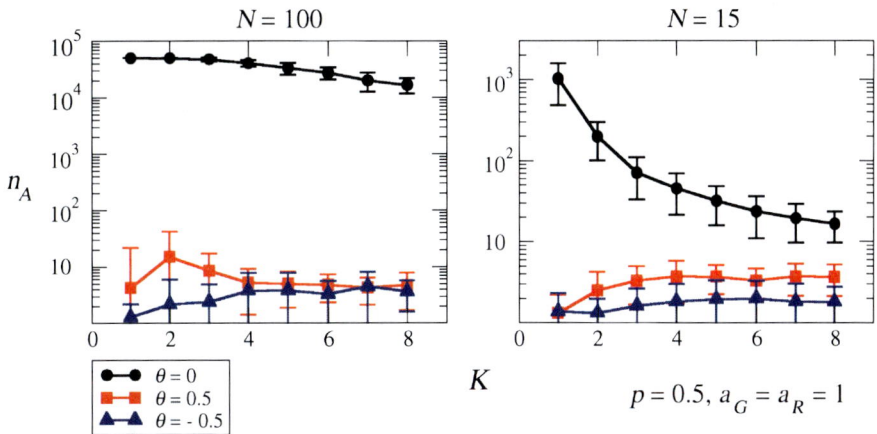

Fig. 7. Average number of attractors as a function of the network connectivity K for RTN's with (a) $N = 100$ and (b) $N = 15$ nodes. In each panel, we used $p = 0.5$, $a_G = a_R = 1$ and $\theta = 0$ (circles), $\theta = 0.5$ (squares) and $\theta = -0.5$ (triangles). For the large networks in (a) we sample the configuration space using $50,000$ randomly chosen initial conditions, whereas in (b) the full configuration space was probed. Both in (a) and (b) each point is the average over ensembles of 50 networks. Note the extremely large number of attractors obtained for $\theta = 0$, especially for moderately small values of K. In particular, for $\theta = 0$ in (a) almost every sampled initial condition leads to a different attractor.

The $\theta = 0$ case presents "anomalous" behavior not only with regard to the number of attractors, but also in the structure of the state space.[10] Fig. 8 shows the largest basin of attraction for three network realizations with $N = 15$, $K = 8$, $p = 0.5$, $a_R = a_G = 1$, and $\theta = 0.5$ (left part) and $\theta = 0$ (right part). For those parameters the networks are in the chaotic regime. It can be seen from this figure that the structure of the largest basin of attraction for the non-integer threshold is characterized by long transients and attractor length. This is similar to what is obtained using RBN's with the same p and the same K. However, for $\theta = 0$ the structure is quite different. Note first that all the attractors are punctual. Although this is not the rule, it is the most probable situation for $\theta = 0$. Additionally, the transients are comparatively short and the whole basins look somehow sparse as compared to the ones on the left part. The long arm-like structures observed in the basins of attraction for $\theta = 0.5$ reflect that the routes to reach the attractor are concentrated in a few number of states. From a biological point of view, these few states can be considered as the "checkpoints" of the differentiation (or metabolic) pathway. Contrary to this, the sparseness observed in the basins of attraction for $\theta = 0$ indicate that the routes to reach the attractor are much more distributed across the state space, and therefore, the existence of "checkpoints" is harder to attain.

---

[10] Here, by "anomalous" we mean with respect to what is observed in standard Kauffman networks.

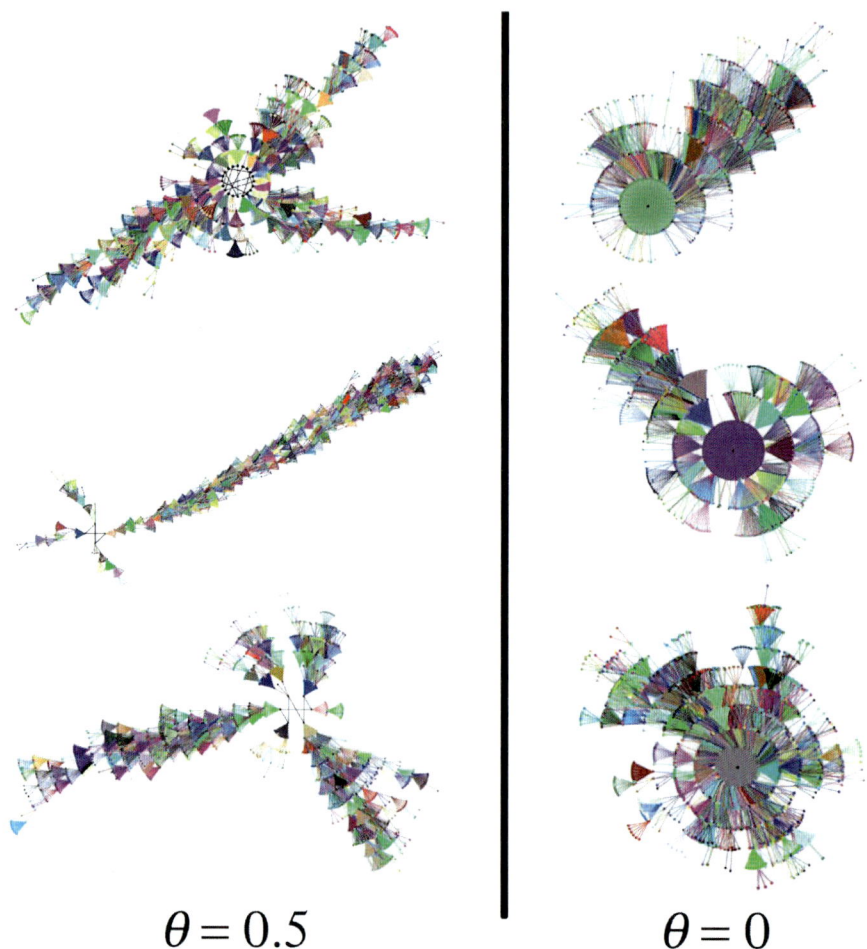

Fig. 8. Structure of the attractor landscape of RTN's with $N = 15$, $K = 8$, $p = 0.5$, $a_R = a_G = 1$, and $\theta = 0.5$ (left) and $\theta = 0$ (right). Each of the basins of attraction shown is the largest one in a given network realization. Note that for $\theta = 0.5$ the attractors have several configurations (black dots at the center of each structure), whereas for $\theta = 0$ all the atractors are punctual (only one dot at the center). Additionally, for $\theta = 0.5$ the long arm-like structures indicate the existence of long transients and a reduced number of configurations which all the routes that go to the attractor have to go through. Contrary to this, for $\theta = 0$ the basins of attraction look more sparse and with shorter transients and more distributed across the configuration space.

5 Summary and Discussion

We have investigated the dynamical properties of random threshold networks (RTN's), which differ from standard Kauffman networks (or random Boolean networks, RBN's) in that the regulation of the state of the nodes is done by means of threshold functions. These networks have been used in the modeling of genetic regulatory networks of real organisms using parameter values that seem biologially meaningful [15,16]. An important characteristic of the threshold network model is that it assumes that gene regulation is an additive process. Namely, that the combined effect of the regulators of a given gene on the state of that gene is just the sum of the positive regulations minus the negative ones. Because of its simplicity, this assumption is very tempting when constructing models of gene regulatory networks. However, there are many examples in real systems showing that gene regulation is combinatorial rather than additive, which means that the effect of some regulators (i.e. whether activatory or inhibitory) depends on the presence or absence of some other regulators.[11] In several cases, these combinatorial processes are represented by Boolean functions that cannot be obtained from threshold functions (for instance, the XOR function), which makes the additivity assumption mentioned above questionable.

Additionally, a deeper analysis of the dynamics of RTN's reveals anomalies inconsistent with the expected behavior of gene regulation models, precisely for the "biologically meaningful" values of the parameters that have been used. Of particular importance is the case of integer threshold values (illustrated here using $\theta = 0$), where the networks typically have an enormous amount of attractors. Also in this case, there is a sharp disagreement between the ensemble properties predicted by the annealed approximation and the ones observed in the numerical simulation using concrete network realizations. It is worth emphasizing that this disagreement happens neither for RTN's with non-integer threshold values nor for RBN's. It is not surprising to find such a disagreement in specific networks constructed in very peculiar ways (as the ones used for the modeling of the cell-cycles). What is surprising is that the typical members of the ensemble, constructed in a completely random way, present such anomalies. In fact, the cell-cycle networks in Refs. [15,16] do exhibit these anomalies, as they have a large number of attractors. However, the authors of that work considered most of these attractors as biologically irrelevant because of their small basins sizes, and neglected them. Nonetheless, from an evolutionary point of view it is not clear whether or not the size of the basin of attraction is relevant, as it is not known whether this parameter is under selective pressure. Actually, in other studies of gene regulatory networks of real organisms, the biologically meaningful attractors of the wild-type organism do not possess the largest basins of attraction, but on the contrary, sometimes they have very tiny basins [12,13].

We computed analytically the phase diagram that determines in which parameter region the network has chaotic, ordered or critical dynamics. Contraty to what happens for RBN's, the phase diagrams obtained for RTN's are always asymmetric with respect to the fraction of positive regulations p. It is only for non-integer positive thresholds (illustrated here for the case $\theta = 0.5$), that the phase diagram looks semi-symmetric, similar to the one obtained for RBN's (see Fig. 2). This may be important because in such a case there is a bigger freedom

---

[11] An common example of this are dual regulators in *E. coli* [41].

to vary p and still remain close to the critical region, especially at low newtork connectivities as the ones reported for networks of real organisms that operate in the critical regime ($K \sim 2$, [34]). Note that the case $\theta = 0.5$ is biologically reasonable not only in terms of the phase diagram, but also because it corresponds to a situation in which at least one positive regulator has to be active in order to activate the target gene. Contrary to this, in the case $\theta = -0.5$ all genes get activated when all of its positive become innactive, which is an artifact of the model rather than a biological observed behavior. Furthermore, the phase transition for $\theta = -0.5$ and $p = 0.5$ occurs at a network connectivity $K = 4$ (seel Fig. 4), which is large compared to the one observed in real networks. Therefore, this case with negative threshold values seems to be inadequate for the study of the theoretical properties of gene regulatory networks. Consequently, some of the conclusions about the evolution of RTN's with negative thresholds might have to be reinterpreted [43].

One important point analyzed in this work was the usefulness of the Derrida map to elucidate the network's dynamical regime. As it is shown in Fig. 6, for RTN's the first steps in the dynamics may indicate that the network is in the ordered regime, while in fact the long-term behavior is chaotic. This occurs when long-term correlations are developed thoughout time, which always happens in RTN's, especially for integer threshold values. For in such a case, the self-regulation implied by the last line in Eq. (2) induces long-term memory in the system. For RBN's these long-term correlations do not exist, and the Derrida map accurately predicts the network's dynamical regime. This is an important aspect that has not been properly taken into account in current work that aims to characterize the network's dynamical regime in real biological networks. Therefore, a more thorough study is necessary in this direction.

Finally, it is important to note that in this work we used the same fixed connectivities, interaction strengths and thresholds for all genes. However, more realistic situations would require assigning these quantities differently to the different genes in the network. For instance, for some genes the inhibitory regulators may be dominant, whereas for other genes the activatory regulators would dominate. Also, genes with integer as well as non-integer threshold values can coexist in the same network. Exploration of these possibilities can reveal dynamical behaviors more consistent with biological systems, which in turn would help to discern the model's characteristics relevant for biological modeling.

Acknowledgements

We thank H. Larralde for fruitful discussions. J.G.T. Zañudo acknowledges CONACyT for a research asistant scholarship. This work was partially supported by PAPIIT-UNAM grants IN112407-3 and IN109210.

References

1. Glass, L., Kauffman, S.A.: The logical analysis of continous, nonlinear biochemical control networks. J. Theor. Biol. 39, 103–129 (1973)
2. Tyson, J.J., Chen, K.C., Novak, B.: Sniffers, buzzers, toggles and blinkers: dynamics of regulatory and signaling pathways in the cell. Curr. Op. Cell Biol. 15, 221–231 (2003)
3. Tyson, J.J., Chen, K.C., Novak, B.: Network dynamics and cell physiology. Nature Rev. Mol. Cell Biol., 908–916 (2001)
4. Bornholdt, S.: Systems biology: Less is more in modeling large genetic networks. Science 310(5747), 449–451 (2005)
5. Albert, R., Othmer, H.G.: ...but no kinetic details are needed. SIAM News 36(10) (December 2003)
6. Wang, R., Albert, R.: Discrete dynamic modeling of cellular signaling networks. In: Johnson, B.M.L., Brand, L. (eds.) Methods in Enzymology 476: Computer Methods, pp. 281–306. Academic Press, London (2009)
7. Albert, R., Toroczkai, Z., Toroczkai, Z.: Boolean modeling of genetic networks. In: Ben-Naim, E., Frauenfelder, H., Toroczkai, Z. (eds.) Complex Networks. Springer, Heidelberg (2004)
8. Aldana-Gonzalez, M., Coppersmith, S., Kadanoff, L.P.: Boolean Dynamics with Random Couplings. In: Kaplan, E., Marsden, J.E., Sreenivasan, K.R. (eds.) Perspectives and Problems in Nonlinear Science. A celebratory volule in honor of Lawrence Sirovich. Springer Applied Mathematical Sciences Series, pp. 23–89 (2003)
9. Kauffman, S.A.: Metabolic stability and epigenesis in randomly constructed genetic nets. J. Theor. Biol. 22, 437–467 (1969)
10. Kauffman, S.A.: The Origins of Order: Self-organization and selection in evolution. Oxford University Press, Oxford (1993)
11. Mendoza, L., Thieffry, D., Alvarez-Buylla, E.R.: Genetic control of flower morphogenesis in Arabidopsis Thaliana: a logical analysis. Bioinformatics 15, 593–606 (1999)
12. Espionza-Soto, C., Padilla-Longoria, P., Alvarez-Buylla, E.R.: A gene regulatory network model for cell-fate determination during Arabidopsis Thaliana flower development that is robust and recovers experimental gene expression profiles. Plant Cell. 16, 2923–2939 (2004)
13. Albert, R., Othmer, H.G.: The topology of the regulatory interactions predicts the expression pattern of the segment polarity genes in Drosophila melanogaster. J. Theor. Biol. 23, 1–18 (2003)
14. Mendoza, L.: A network model for the control of the differentiation process in Th cells. BioSystems 84, 101–114 (2006)
15. Li, F., Long, T., Lu, Y., Ouyang, Q., Tang, C.: The yeast cell-cycle network is robustly designed. Proc. Natl. Acad. Sci. USA 101, 4781–4786 (2004)
16. Davidich, M.I., Bornholdt, S.: Boolean network model predicts cell cycle sequence of fission yeast. PLoS ONE 3(2), e1672 (2008)
17. Perissi, V., Jepsen, K., Glass, C.K., Rosenfeld, M.G.: Deconstructing repression: evolving models of co-repressor action. Nature Reviews Genetics 11, 109–123 (2010)
18. McCulloch, W., Pitts, W.: A logical calculus of the ideas immanent in nervous activity. Bull. Math. Biophys. 7, 115–133 (1943)

19. Hertz, J., Krogh, A., Palmer, R.G.: Introduction to the Theory of Neural Computation. Santa Fe Institute Studies in the Science of Complexity. Addison-Wesley, Reading (1991)
20. Bornholdt, S.: Boolean network models of cellular regulation: prospects and limitations. J. R. Soc. Interface 5, S85–S94 (2008)
21. Kürten, K.E.: Critical phenomena in model neural networks. Phys. Lett. A 129, 157–160 (1988)
22. Rohlf, T., Bornholdt, S.: Criticality in random threshold networks: annealed approximation and beyond. Physica A 310, 245–259 (2002)
23. Aldana, M., Larralde, H.: Phase transitions in scale-free neural networks: Departure for the standard mean-field universality class. Phys. Rev. E 70, 066130 (2004)
24. Rohlf, T.: Critical line in random threshold networks with inhomogeneous thresholds. Phys. Rev. E 78, 066118 (2008)
25. Kürten, K.E.: Correspondance between neural threshold networks and Kauffman Boolean cellular automata. J. Phys. A 21, L615-L619 (1988)
26. Derrida, B.: Dynamical phase transition in nonsymmetric spin glasses. J. Phys. A: Math. Gen. 20, L721-L725 (1987)
27. Szejka, A., Mihaljev, T., Drossel, B.: The phase diagram of random threshold networks. New Journal of Physics 10, 063009 (2008)
28. Derrida, B., Pomeau, Y.: Random networks of automata: a simple annealed approximation. Europhys. Lett. 1(2), 45–49 (1986)
29. Moreira, A.A., Amaral, L.A.N.: Canalyzing Kauffman networks: Nonergodicity and its effect on their critical behavior. Phys. Rev. Lett. 94, 0218702 (2005)
30. Serra, R., Villani, M., Graudenzi, A., Kauffman, S.A.: Why a simple model of genetic regulatory networks describes the distribution of avalanches in gene expression data. J. Theor. Biol. 246, 449–460 (2007)
31. Serra, R., Villani, M., Semeria, A.: Genetic network models and statistical properties of gene expression data in knock-out experiments. J. Theor. Biol. 227, 149–157 (2004)
32. Shmulevich, I., Kauffman, S., Aldana, M.: Eukaryotic cells are dynamically ordered or critical but not chaotic. Proc. Natl. Acad. Sci. USA 102, 13439–13444 (2005)
33. Nykter, M., Price, N.D., Aldana, M., Ramley, S.A., Kauffman, S.A., Hood, L.E., Yli-Harja, O., Shmulevich, I.: Gene expression dynamics in the macrphage exhibit criticality. Proc. Natl. Acad. Sci. USA 105(6), 1897–1900 (2008)
34. Balleza, E., Alvarez-Buylla, E.R., Chaos, A., Kauffman, S., Shmulevich, I., Aldana, M.: Critical dynamics in genetic regulatory networks: Examples from four kingdoms. PLoS ONE 3(6), e2456 (2008)
35. Derrida, B., Weisbuch, G.: Evolution of overlaps between configurations in random Boolean networks. J. Phys. (Paris) 47, 1297–1303 (1986)
36. Aldana, M.: Boolean dynamics of networks with scale-free topology. Physica D 185, 45–66 (2003)
37. Kesseli, J., Rämö, P., Yli-Harja, O.: Iterated maps for annealed Boolean networks. Phys. Rev. E 74, 046104 (2006)
38. Greil, F., Drossel, B.: Kauffman networks with threshold functions. Eur. Phys. J. B 57, 109–113 (2007)
39. Kauffman, S.A.: Requirements for evolvability in complex systems: orderly dynamics and frozen components. Physica D 42(1-3), 135–152 (1990)
40. Aldana, M., Balleza, E., Kauffman, S.A., Resendiz, O.: Robustness and evolvability in genetic regulatory networks. J. Theor. Biol. 245, 433–448 (2007)

41. Gama-Castro1, S., Jiménez-Jacinto, V., Peralta-Gil, M., Santos-Zavaleta, A., Peñaloza-Spinola, M.I., Contreras-Moreira, B., Segura-Salazar, J., Muñiz-Rascado, L., Martínez-Flores, I., Salgado, H., Bonavides-Martínez, C., Abreu-Goodger, C., Rodríguez-Penagos, C., Miranda-Ríos, J., Morett, E., Merino, E., Huerta, A.M., Treviño-Quintanilla, L., Collado-Vides, J.: RegulonDB (version 6.0): gene regulation model of Escherichia coli K-12 beyond transcription, active (experimental) annotated promoters and Textpresso navigation. Nucleic Acids Research 36, D120-D124 (2008)
42. Braunewell, S., Bornholdt, S.: Superstability of the yeast cell-cycle dynamics: Ensuring causality in the presence of biochemical stochasticity. J. Theor. Biol. 245, 638–643 (2007)
43. Szejka, A., Drossel, B.: Evolution of Boolean networks under selection for a robust response to external inputs yields an extensive neutral space. Phys. Rev. E 81, 021908 (2010)

Appendix

Appendix A: Derivation of the Map $b(t+1) = B(b(t))$

We first remember that $b(t)$ represents the fraction of active nodes ($\sigma_i = 1$) for a given network configuration at time t. Using the annealed approximation, the evolution of $b(t)$ depends only on the previous state and can thus be given by the map B, which relates the number of active nodes after two consecutive time steps.

To explicitly obtain B let us consider the following. Since in the annealed approximation we assume statistical independence between the nodes, the fraction of active elements $b(t)$ can actually be considered as the probability that an arbitrary node σ_i is active at time t. Therefore, $b(t+1)$ corresponds to the probability that a node is active after one time step. From the dynamical equation for the nodes, Eq. (2), it is apparent that there are only 2 ways in which this can happen: Either the sum $\sum_j a_{n,j}\sigma_{n_j}(t)$ was larger than θ or it was equal to θ. In this last case we additionally need the node itself to be active at time t so that it is still active at time $t+1$, which happens with probability $b(t)$. If we denote $p_+(b(t))$ as the probability that $\sum_j a_{n,j}\sigma_{n_j}(t) > \theta$ and $p_0(b(t))$ as the probability that $\sum_j a_{n,j}\sigma_{n_j}(t) = \theta$, then B must be the sum of the probabilities of these two events:

$$b(t+1) = B(b(t)) = p_+(b(t)) + b(t) \cdot p_0(b(t)), \qquad (1)$$

This corresponds to Eq. (5a) in the main text. The full expression for p_0 and p_+ are derived in the next section, Appendix B.

Appendix B: Derivation of $p_0(b)$ and $p_+(b)$

To derive these expressions we use the mean-field method from Ref. [23]. Let us denote $P_\Sigma(y)$ as the probability distribution function of the sum $\xi_i = \sum_{j=1}^{k_i} a_{i,j}\sigma_{i_j}$

in Eq. (2) of a node σ_i with k_i inputs. The probability that $\xi_i = 0$ and $\xi_i > 0$ are then given, respectively, by

$$p_+(b, k_i) = \lim_{\epsilon \to 0} \int_{\theta+\epsilon}^{\infty} P_\Sigma(y)\, dy \tag{1}$$

$$p_0(b, k_i) = \lim_{\epsilon \to 0} \int_{\theta-\epsilon}^{\theta+\epsilon} P_\Sigma(y)\, dy. \tag{2}$$

The weights $a_{i,j}$ can be considered as random variables which take the value a_G with probability p and $-a_R$ with probability $q = 1 - p$, as defined in Sec.2. Using this and denoting b as the probability that a node is active, we can consider the $\xi_{ij} = a_{i,j}\sigma_{i_j}$ as random variables which can take the values 0 with probability $1 - b$, a_G with probability bp and $-a_R$ with probability bq, that is

$$P_\xi(x) = (1-b)\delta(x) + bp\,\delta(x - a_G) + bq\,\delta(x + a_R), \tag{3}$$

Using the statistical independence assumption of the annealed approximation, each ξ_{ij} in the sum $\xi_i = \sum_{j=1}^{k_i} \xi_{ij}$ is an independent random variable with probability distribution $P_\xi(x)$. Because of this, ξ_i is the sum of k_i independent random variables, and thus $P_\Sigma(y)$ must be the k_i-fold convolution of P_ξ:

$$P_\Sigma(y) = \underbrace{P_\xi * P_\xi * \cdots * P_\xi(y)}_{k_i \text{ times}}. \tag{4}$$

Taking the Fourier transform of the above equation we get

$$\hat{P}_\Sigma(\omega) = \left[\hat{P}_\xi(\omega)\right]^{k_i}, \tag{5}$$

where $\hat{P}_\xi = (1-b) + bpe^{-i\omega a_G} + bqe^{i\omega a_R}$. Thus, $P_\Sigma(y)$ is obtained by taking the inverse transform of the last equation. Using the binomial theorem twice we get

$$P_\Sigma(y) = \frac{1}{2\pi} \int_{-\infty}^{\infty} e^{i\omega y} \left[\hat{P}_\xi(\omega)\right]^{k_i} d\omega$$

$$= \frac{1}{2\pi} \sum_{i=0}^{k_i} \int_{-\infty}^{\infty} e^{i\omega y} \binom{k_i}{i} (1-b)^i b^{k_i - i} \left(pe^{-i\omega a_G} + qe^{i\omega a_R}\right)^{k_i - i} d\omega$$

$$= \frac{1}{2\pi} \sum_{i=0}^{k_i} \sum_{l=0}^{k_i - i} \binom{k_i}{i}\binom{k_i - i}{l} (1-b)^i b^{k_i - i} p^l q^{k_i - i - l} \tag{6}$$

$$\times \int_{-\infty}^{\infty} e^{i\omega y} e^{-i\omega a_G l} e^{i\omega a_R (k_i - i - l)}\, d\omega$$

$$= \sum_{i=0}^{k_i} \sum_{l=0}^{k_i - i} \binom{k_i}{i}\binom{k_i - i}{l} (1-b)^i b^{k_i - i} p^l q^{k_i - i - l} \delta\left[a_G l - a_R(k_i - i - l) - y\right] \tag{7}$$

If we substitute this last result into Eqs. (1) and (2) we find

$$p_+(b, k_i) = \lim_{\epsilon \to 0} \int_{\theta+\epsilon}^{\infty} P_\Sigma(y) \, dy$$

$$= \sum_{i=0}^{k_i} \binom{k_i}{i} (1-b)^i b^{k_i-i}$$

$$\times \lim_{\epsilon \to 0} \int_{\theta+\epsilon}^{\infty} \left\{ \sum_{l=0}^{k_i-i} \binom{k_i-i}{l} p^l q^{k_i-i-l} \delta[a_G l - a_R(k_i-i-l) - y] \right\} dy$$

$$= \sum_{i=0}^{k_i} \binom{k_i}{i} (1-b)^i b^{k_i-i} \sum_{l=l_i}^{k_i-i} \binom{k_i-i}{l} p^l q^{k_i-i-l}, \quad (8)$$

where $l_i = \dfrac{(k_i - i)a_G + \theta}{a_G + a_R} + 1$

$$p_0(b, k_i) = \lim_{\epsilon \to 0} \int_{\theta-\epsilon}^{\theta+\epsilon} P_\Sigma(y) \, dy$$

$$= \sum_{i=0}^{k_i} \binom{k_i}{i} (1-b)^i b^{k_i-i}$$

$$\times \lim_{\epsilon \to 0} \int_{\theta-\epsilon}^{\theta+\epsilon} \left\{ \sum_{l=0}^{k_i-i} \binom{k_i-i}{l} p^l q^{k_i-i-l} \delta[a_G l - a_R(k_i-i-l) - y] \right\} dy$$

$$= \sum_{i=0}^{k_i} \binom{k_i}{i} (1-b)^i b^{k_i-i} \sum_{l=0}^{k_i-i} \binom{k_i-i}{l} p^l q^{k_i-i-l} \delta_{a_G l, a_R(k_i-i-l)+\theta} \quad (9)$$

where in Eq. (8) the minimum value of $l = l_i$ was chosen so that the argument of the Dirac delta function is always above θ, as specified by the limit. Similarly in Eq. (8) it is chosen so that it is exactly equal to θ. Finally, since the probability distribution of k_i is given by $P_{in}(k)$ we have that

$$p_+(b, k_i) = \sum_{k_i=1}^{\infty} P_{in}(k_i) \, p_+(b, k_i)$$

$$= \sum_{k_i=1}^{\infty} P_{in}(k_i) \sum_{i=0}^{k_i} \binom{k_i}{i} (1-b)^i b^{k_i-i} \sum_{l=l_i}^{k_i-i} \binom{k_i-i}{l} p^l q^{k_i-i-l}, \quad (10)$$

$$p_0(b, k_i) = \sum_{k_i=1}^{\infty} P_{in}(k_i) \, p_0(b, k_i)$$

$$= \sum_{k_i=1}^{\infty} P_{in}(k_i) \sum_{i=0}^{k_i} \binom{k_i}{i} (1-b)^i b^{k_i-i} \sum_{l=0}^{k_i-i} \binom{k_i-i}{l} p^l q^{k_i-i-l}$$

$$\times \delta_{a_G l, a_R(k_i-i-l)+\theta}. \quad (11)$$

which correspond, respectively, to Eqs. (5b) and (5c).

Appendix C: Derivation of $I^{(k_d)}$ for Boolean Threshold Networks

We first remember the definition of $I^{(k_d)}$, the influence of k_d variables. Denoting Σ_t and $\widetilde{\Sigma}_t$ as two configurations in which k_d of the inputs of an arbitrary node σ_i are different, $I^{(k_d)}$ is the probability that, after a time step, the node σ_i in the new configurations, Σ_{t+1} and $\widetilde{\Sigma}_{t+1}$, are different from each other.

Consider the average over all possible active/inactive and activatory/inhibitory configurations of the inputs of an arbitrary node σ_i. If the node has k_i inputs, activator probability p and a fraction of active nodes b, then this average is given by

$$\langle X \rangle_{IC} = \sum_{i=0}^{k_i} \binom{k_i}{i} p^i q^{k_i-i} \sum_{l=0}^{i} \sum_{m=0}^{k_i-i} \binom{i}{l}\binom{k_i-i}{m} b^{l+m}(1-b)^{k_i-l-m} X. \quad (1)$$

Here $p^i q^{k_i-i}$ ($q = 1-p$) is the probability that for a given input configuration with k_i regulators there are i activatory interactions and $k_i - i$ inhibitory ones, which can be chosen in $\binom{k_i}{i}$ possible ways. $b^{l+m}(1-b)^{k_i-l-m}$ is the probability that there are l active activatory inputs and m active inhibitory ones, which can be arranged in $\binom{i}{l}\binom{k_i-i}{m}$ different ways. Since $I^{(k_d)}$ is the probability for one arbitrary node (regardless of the number of active or inactive inputs), we have to compute the average over all possible input configurations of $\mathcal{I}(k_i, k_d, i, l, m)$, which is the probability that a damage spreads when k_d of the input elements are damaged given that this configuration has i activatory inputs and $k_i - i$ inhibitory ones, which in turn have l and m active/inactive input nodes, respectively.

To find $\mathcal{I}(k_i, k_d, i, l, m)$ we need to consider all possible ways in which the damaged nodes can be arranged. There are l active activatory input nodes and m active inhibitory ones, and thus, $i-l$ inactive activatory elements and k_i-i-m inactive inhibitory ones. Therefore, we may have u damaged active activatory inputs, v damaged active inhibitory ones, w damage inactive activatory ones and $z = k_d - u - v - w$ damaged inactive ones. Since there are $\binom{l}{u}\binom{m}{v}\binom{i-l}{w}\binom{k_i-i-m}{k_d-u-v-w}$ possible ways in which damage can be distributed, and given that there are $\binom{k_i}{k_d}$ total ways in which the damaged nodes can be arranged, then the probability for each value of u, v, w is given by a multivariate hypergeometric distribution

$$\Pr(u,v,w) = \frac{\binom{l}{u}\binom{m}{v}\binom{i-l}{w}\binom{k_i-i-m}{k_d-u-v-w}}{\binom{k_i}{k_d}} \quad \begin{matrix} u = u_0, \ldots, u_f \\ v = v_0, \ldots, v_f \\ w = w_0, \ldots, w_f \end{matrix}, \quad (2)$$

where

$$\begin{aligned}
u_0 &= \max(0, k_d + l - k_i), & u_f &= \min(l, k_d) \\
v_0 &= \max(0, k_d - u - (k_i - l - m)), & v_f &= \min(l, k_d - u) \\
w_0 &= \max(0, k_d - u - v - (k_i - l - m - i + l)), & w_f &= \min(l, k_d - u - v).
\end{aligned} \quad (3)$$

Finally, we need to consider all possible ways in which a damage can actually make the state of σ_i in the damage and the undamaged configuration different at time $t+1$. From the definitions of l and m it follows that, before damage, there are l active activatory input nodes and m active inhibitory ones. After damage, using the definitions of u, v, w and z, there will be $l - u + w$ active activatory input elements and $m - v + z$ active inhibitory ones. Using this information and Eq. (2), it is clear that the damage can spread in 3 different ways:

i) If in the damaged configuration the sum is above the threshold ($a_G(l - u + w) - a_R(m - v + z) > \theta$), then the state of the nodes can be different if, without the damage the sum is either below the threshold ($a_G l - a_R m < \theta$) or exactly at the threshold ($a_G l - a_R m = \theta$) but with the condition that it was inactive at the time-step before (so that they are different in the next time step), which happens with probability $1 - b$.

$$P_1 = H\left(a_G(l - u + w) - a_R(m - v + z) - \theta\right) \cdot \left[(1 - b)\delta_{a_G l - a_R m, \theta} + H(a_R m + \theta - a_G l)\right]. \quad (4a)$$

ii) The second possiblity is that in the damaged configuration the sum is just at the threshold ($a_G(l - u + w) - a_R(m - v + z) = \theta$). In that case damage can spread if: (a) before the damage the sum is above the threshold ($a_G l - a_R m > \theta$) but only if it was inactive before (with probability $1 - b$); (b) if before the damage the sum is below the threshold ($a_G l - a_R m < \theta$) but only if it was active on the step before (with probability b); and (c) if before the damage the sum is again exactly at the threshold ($a_G l - a_R m = \theta$) but only if both configurations where damaged before (with probability $h(t)$).

$$P_2 = \delta_{a_G(l-u+w), a_R(m-v+z)+\theta} \left[h(t)\delta_{a_G l - a_R m, \theta} + b\, H(a_R m + \theta - a_G l) + (1 - b)H(a_G l - a_R m - \theta)\right]. \quad (4b)$$

iii) In the third possibility the damaged configuration has its sum below the threshold ($a_G(l-u+w)-a_R(m-v+z) < \theta$) so the state of the nodes will differ if before the damage the sum is either above the threshold ($a_G l - a_R m > \theta$) or exactly at the threshold ($a_G l - a_R m = \theta$) but only if it was active at the time-step before, which happens with probability b.

$$P_3 = H\left(a_R(m - v + z) + \theta - a_G(l - u + w)\right) \cdot \left[b\, \delta_{a_G l - a_R m, \theta} + H(a_G l - a_R m - \theta)\right]. \quad (4c)$$

Since all three cases can make damage spread, then the total probability for damage spreading $\mathcal{I}$ is the sum of these three possiblilities: Eqs. (4a), (4b) and (4c), averaged over all possible arrangements of damaged configurations, Eq. (3),

$$\mathcal{I} = \sum_{u=u_0}^{u_f} \sum_{v=v_0}^{v_f} \sum_{w=w_0}^{w_f} Pr(u,v,w) \cdot (P_3 + P_2 + P_3)$$

$$= \sum_{u=u_0}^{u_f} \sum_{v=v_0}^{v_f} \sum_{w=w_0}^{w_f} \frac{\binom{l}{u}\binom{m}{v}\binom{i-l}{w}\binom{k_i-i-m}{k_d-u-v-w}}{\binom{k_i}{k_d}}$$

$$\times \{H\left(a_G(l-u+w) - a_R(m-v+z) - \theta\right) \cdot [(1-b_\infty)\delta_{a_G l - a_R m, \theta}$$
$$+ H(a_R m + \theta - a_G l)] + \delta_{a_G(l-u+w), a_R(m-v+z)+\theta} \cdot [h(t)\delta_{a_G l - a_R m, \theta}$$
$$+ b_\infty H(a_R m + \theta - a_G l) + (1 - b_\infty)H(a_G l - a_R m - \theta)]$$
$$+ H\left(a_R(m-v+z) + \theta - a_G(l-u+w)\right) \cdot [b_\infty \delta_{a_G l - a_R m, \theta}$$
$$+ H(a_G l - a_R m - \theta)]\} . \quad (5)$$

Finally, averaging this damage spread for a given input configuration $\mathcal{I}(k_i, k_d, i, l, m)$ over all possible input configurations using Eq. (1), we get the average influence of k_d variables:

$$I^{(k_d)} = \langle \mathcal{I}(k_i, k_d, i, l, m) \rangle_{IC}$$
$$= \sum_{i=0}^{k_i} \binom{k_i}{i} p^i q^{k_i - i} \sum_{l=0}^{i} \sum_{m=0}^{k_i - i} \binom{i}{l} \binom{k_i - i}{m} b^{l+m} (1-b)^{k_i - l - m}$$
$$\times \mathcal{I}(k_i, k_d, i, l, m). \quad (6)$$

Eqs. (5) and (6) with $b = b_\infty$ correspond to the formulas (7) and (6), respectively, in the main text.

Appendix D: Derivation of S and $I^{(0)}$

As discussed in Sec. 4.2, $I^{(0)}$ is the probability that, for an arbitrary node σ_i, damage spreads at the next time step when none of its input elements are different between the two initial configurations, $\Sigma_t, \widetilde{\Sigma}_t$. Because of the possibility of having this sum giving exactly the threshold value θ, $I^{(0)}$ is zero only for noninteger thresholds.

The case for integer values of θ can be obtained from Eqs. (6) and (7). However, from Eq. (2) we can see that the only way for a damage to spread when none of the input elements are different is by having $\sum_j a_{n,j} \sigma_{n_j}(t) = \theta$. Thus, $I^{(0)}$ must correspond to the probability p_0 that the sum gives exactly the threshold value, Eq. (5c). Since we need to consider the nonergodicity of the system, we use the final fraction of activatory nodes b_∞ as the value of b. Now, for damage to spread not only does the sum need to be at the threshold, but also the two nodes must be different initially, otherwise they would be the same in the next time-step and the damage would not spread. Since $h(t)$ can be considered as the probability that two arbitrary nodes are different, then $I^{(0)}$ must be the multiplication of both probabilities

$$I^{(0)} = p_0(b_\infty) \cdot h(t) \quad \text{with } \theta \in \mathbb{Z}, \quad (1)$$

Taking this last result into consideration we can now calculate the sensitivity S of RTN's. Using (8) in (4) we get

$$S = \sum_{k_i=1}^{\infty} P_{in}(k_i) \left[\frac{dI^{(0)}}{dh} + k_i(I^{(1)} - I^{(0)}) \right]\bigg|_{h=0}. \quad (2)$$

From Eq. (7) it is apparent that $I^{(1)}$ does not actually depend on h. This happens because the middle term of Eq. (7), which is the only one with a h dependance, can never be nonzero for the case of $k_d = 1$. This is because $I^{(1)}$ refers to the case where after k_d changes of an arbitrary input, the summation of both configurations is still at the threshold, which cannot happen when only 1 input is changed. With this information and using Eq. (1) in (2) we finally find

$$S = \sum_{k_i=1}^{\infty} P_{in}(k_i) \left[p_0(b_\infty) + k_i \left(I^{(1)} - I^{(0)} \right) p_0(b_\infty) \cdot h \right]\bigg|_{h=0}$$

$$= \sum_{k_i=1}^{\infty} P_{in}(k_i) \left(p_0(b_\infty) + k_i I^{(1)} \right)$$

$$= p_0(b_\infty) + \sum_{k_i=1}^{\infty} P_{in}(k_i) k_i I^{(1)}. \quad (3)$$

Eqs. (1) and (3) correspond, respectively, to Eqs. (9) and (10) in the main text.

Appendix E: Derivation of S_0 for $p = 0.5$, $a_G = a_R = 1$, $\theta = 0$ and $\theta = \pm 0.5$

We first remember from Sec. 4.1 that S_0, the uncorrelated network sensitivity, is the average number of nodes by which two configurations differ after one time step if they initially differed in only one element:

$$S_0 = N \left\langle d\left(\Sigma_1, \tilde{\Sigma}_1\right) \right\rangle, \quad \text{with} \quad d\left(\Sigma_0, \tilde{\Sigma}_0\right) = \frac{1}{N}.$$

Because of the thermodynamic limit assumed in the annealed approximation, S_0 should correspond to the sensitivity of the network given in Eq. (10) with $b_\infty = b_0 = 0.5$. This last choice of b is a consequence of the initial configurations being chosen randomly. In what follows we consider consider the case ni which $P_{in}(k) = \delta_{k,K}$ with $p = 0.5$, $a_G = a_R = 1$, $\theta = 0$ and $\theta = \pm 0.5$.

E.1 S_0 for $p = 0.5$, $a_G = a_R = 1$ and $\theta = 0$

Using Eqs. (9) and (10) for the integer threshold $\theta = 0$, we have

$$S_0 = p_0(b_0) + K I^{(1)}. \quad (1)$$

where from Eqs. (7) and (5c)

$$p_0(b_0) = \frac{1}{2^{2K}} \sum_{i=0}^{K} \binom{K}{i} 2^i \sum_{l=0}^{K-i} \binom{K-i}{l} \delta_{l,K-i-l}, \qquad (2)$$

$$\mathcal{I}^{(1)} = \frac{1}{2^{2K}} \sum_{i=0}^{K} \binom{K}{i} \sum_{l=0}^{i} \sum_{m=0}^{K-i} \binom{i}{l}\binom{K-i}{m} \mathcal{I}(a_G=1, a_R=1, \theta=0, b=0.5). \qquad (3)$$

For $\mathcal{I}(a_G = 1, a_R = 1, \theta = 0, b = 0.5)$ we consider the following. Since we have an integer threshold $\theta = 0$ and we are looking for the influence of 1 variable, the input elements of the damaged and the undamaged configurations differ only by one node. Additionally, since the positive and negative weights have equal strenght, the only possible way for the sum to change sign is if: (a) without damage the node is exactly at the threshold (the Kronecker delta terms in Eqs. 4a and Eq. 4c), or (b) if after the damage it its exactly at the threshold (the first Kronecker Delta term in Eq. 4b). Since by damaging a single node we are not able to have the node again at the threshold because of the weights, the term with $h(t)$ in Eq. 4b cannot be attained. Finally, since $k_d = 1$, the only possible values for u, v, w and $z = k_d - u - v - w$ are 1 on one of them and 0 on the rest. Using all this information and Eq. 5 we have

$$\mathcal{I} = \frac{1}{2} \sum_{u=u_0}^{u_f} \sum_{v=v_0}^{v_f} \sum_{w=w_0}^{w_f} Pr(u,v,w) \{H(l-u+w-m+v-z)\delta_{l,m}$$
$$+\delta_{l-u+w, m-v+z}[H(m-l) + H(l-m)] + H(m-v+z-l+u-w)\delta_{l,m}\}$$
$$= \frac{1}{2}\left\{ \left[\frac{m}{K} + \frac{i-l}{K}\right]\delta_{l,m} + \delta_{l,m-1}\left[\frac{m}{K} + \frac{i-l}{K}\right]\right.$$
$$\left. +\delta_{l-1,m}\left[\frac{l}{K} + \frac{K-i-m}{K}\right] + \left[\frac{l}{K} + \frac{K-i-m}{K}\right]\delta_{l,m}\right\}$$
$$= \frac{1}{2K}\left[\delta_{m,l-1}(K-i+1) + \delta_{m,l+1}(i+1)\right] + \frac{1}{2}\delta_{l,m} \qquad (4)$$

E.1.1 $p_0(b_0)$ To get the result from Eq. (2) we use the finite Laplace Transform method. Let us define a generating function

$$g(z) = \sum_{k=0}^{\infty} \frac{c_k}{k!} z^k \qquad (5)$$

with

$$c_k = 2^{2k} p_0(b_\infty)|_{K=k} = \sum_{i=0}^{k} \binom{k}{i} 2^i f(k-i), \qquad (6)$$

$$f(k-i) = \sum_{l=0}^{k-i} \binom{k-i}{l} \delta_{l,k-i-l}. \qquad (7)$$

Using Eq. (5) and the definition of c_k given in Eq. (6), we obtain

$$g(z) = \sum_{k=0}^{\infty} \frac{c_k}{k!} z^k$$

$$= \sum_{k=0}^{\infty} \sum_{i=0}^{k} \frac{2^i}{i!(k-i)!} z^k f(k-i).$$

With the change of variable $u = k - i$ and inverting the order of summation we get

$$g(z) = \sum_{k=0}^{\infty} \sum_{i=0}^{k} \frac{(2z)^i}{i!(k-i)!} z^{k-i} f(k-i)$$

$$= \sum_{i=0}^{\infty} \sum_{u=0}^{\infty} \frac{(2z)^i}{i! u!} z^u f(u)$$

$$= \left[\sum_{i=0}^{\infty} \frac{(2z)^i}{i!} \right] \left[\sum_{u=0}^{\infty} \frac{z^u}{u!} f(u) \right].$$

Substituting into this last result the value of $f(u)$ given in Eq. (7), making the change of variable $v = u - l$, and again exchanging the sums we get

$$g(z) = e^{2z} \left[\sum_{u=0}^{\infty} \sum_{l=0}^{u} \frac{z^u}{(u-l)! l!} \delta_{l,u-l} \right]$$

$$= e^{2z} \left[\sum_{l=0}^{\infty} \sum_{v=0}^{\infty} \frac{z^{2l}}{l! l!} \right]$$

$$= e^{2z} I_0(2z), \tag{8}$$

where I_0 is the modified Bessel function of the first kind and where we used the series representation of the exponential function and $I_0(z) = \sum_{i=0}^{\infty} (z/2)^{2i} / (i!)^2$. Using the integral representation of $I_0(z)$:

$$I_0(z) = \frac{1}{2\pi} \int_{-\pi}^{\pi} e^{z \cos \theta} d\theta,$$

and the change of variable $\phi = 2\theta$ in Eq. (8) we get

$$g(z) = \frac{1}{2\pi} \int_{-\pi}^{\pi} e^{2z(\cos \theta + 1)} d\theta$$

$$= \frac{1}{4\pi} \int_{-2\pi}^{2\pi} e^{4z \cos^2 \phi} d\phi$$

$$= \sum_{k=0}^{\infty} \frac{z^k}{k!} \left[\frac{4^k}{4\pi} \int_{-2\pi}^{2\pi} \cos^{2k} \phi \, d\phi \right]. \tag{9}$$

Comparing this last result with Eq. (5) and using the formula

$$\int_0^{\pi/2} \cos^{2k} \phi \, d\phi = \frac{(2k-1)!!}{(2k)!!} = \frac{1}{2^{2k}} \binom{2k}{k},$$

we finally find

$$\begin{aligned} c_k &= \frac{4^k}{4\pi} \int_{-2\pi}^{2\pi} \cos^{2k} \phi \, d\phi \\ &= 4^k \frac{2}{\pi} \int_0^{\pi/2} \cos^{2k} \phi \, d\phi \\ &= \binom{2k}{k}, \end{aligned} \quad (10)$$

which, using Eq. (6), gives us the first term of Eq. (1):

$$p_0(b_0) = \frac{1}{2^{2K}} \binom{2K}{K}. \quad (11)$$

E.1.2 $I^{(1)}$ From Eqs. (3) and (4) we define

$$I^{(1)} = \frac{1}{2^{2K}} (f_1 + f_2 + f_3), \quad (12)$$

where

$$f_1 = \frac{1}{2K} \sum_{i=0}^{K} \binom{K}{i} \sum_{l=0}^{i} \sum_{m=0}^{K-i} \binom{i}{l} \binom{K-i}{m} \delta_{m,l-1} (K-i+1), \quad (13)$$

$$f_2 = \frac{1}{2K} \sum_{i=0}^{K} \binom{K}{i} \sum_{l=0}^{i} \sum_{m=0}^{K-i} \binom{i}{l} \binom{K-i}{m} \delta_{m,l+1} (i+1), \quad (14)$$

$$f_3 = \frac{1}{2} \sum_{i=0}^{K} \binom{K}{i} \sum_{l=0}^{i} \sum_{m=0}^{K-i} \binom{i}{l} \binom{K-i}{m} \delta_{m,l}. \quad (15)$$

To reduce these expressions we will use Vandermonde's identity and the mean value of the hypergeometric function

$$\binom{m+n}{r} = \sum_{k=0}^{r} \binom{m}{k} \binom{n}{r-k} \quad (16)$$

$$\sum_{k=0}^{r} k \binom{m}{k} \binom{n}{r-k} = \frac{rm}{m+n} \binom{m+n}{r}. \quad (17)$$

Using these expressions in f_1 with the change of variables $l' = l-1$ and $i' = i-1$

$$f_1 = \frac{1}{2K} \sum_{i=0}^{K} \binom{K}{i}(K-i+1) \sum_{l=0}^{i} \binom{i}{l}\binom{K-i}{l-1}$$

$$= \frac{1}{2K} \sum_{i=0}^{K} \binom{K}{i}(K-i+1) \sum_{l'=0}^{i-1} \binom{i}{i-1-l'}\binom{K-i}{l'}$$

$$= \frac{1}{2K} \sum_{i=0}^{K} (K-i+1) \binom{K}{i}\binom{K}{i-1}$$

$$= \frac{1}{2K} \sum_{i'=0}^{K-1} (K-i') \binom{K}{K-1-i'}\binom{K}{i'}$$

$$= \frac{K+1}{4K} \binom{2K}{K-1} = \frac{1}{4}\binom{2K}{K}. \qquad (18)$$

Doing something similar for f_2 with the change of variable $m' = m - 1$

$$f_2 = \frac{1}{2K} \sum_{i=0}^{K} \binom{K}{i}(i+1) \sum_{m=0}^{K-i} \binom{i}{m-1}\binom{K-i}{m}$$

$$= \frac{1}{2K} \sum_{i=0}^{K} \binom{K}{i}(i+1) \sum_{m'=0}^{K-i-1} \binom{i}{m'}\binom{K-i}{K-i-1-m'}$$

$$= \frac{1}{2K} \sum_{i=0}^{K} (i+1) \binom{K}{i}\binom{K}{K-1-i}$$

$$= \frac{1}{2K} \sum_{i=0}^{K-1} (i+1) \binom{K}{i}\binom{K}{K-1-i}$$

$$= \frac{K+1}{4K}\binom{2K}{K-1} = \frac{1}{4}\binom{2K}{K}. \qquad (19)$$

For f_3 we just need to use Eq. (16) with $m = n = K$

$$f_3 = \frac{1}{2} \sum_{i=0}^{K} \binom{K}{i} \sum_{l=0}^{i} \binom{i}{l}\binom{K-i}{l}$$

$$= \frac{1}{2} \sum_{i=0}^{K} \binom{K}{i} \sum_{l=0}^{i} \binom{i}{i-l}\binom{K-i}{l}$$

$$= \frac{1}{2} \sum_{i=0}^{K} \binom{K}{i}\binom{K}{i}$$

$$= \frac{1}{2} \sum_{i=0}^{K} \binom{K}{i}\binom{K}{K-i} = \frac{1}{2}\binom{2K}{K}. \qquad (20)$$

Using Eqs. (18), (19) and (20) in (12) we find the second part of Eq. (1):

$$I^{(1)} = \frac{1}{2^{2K}} \binom{2K}{K}. \tag{21}$$

We finally get the first part of Eq. (12) using Eqs. (11) and (21) in Eq. (1), which gives

$$S_0(p = 0.5, a_G = 1, a_R = 1, \theta = 0) = \frac{K+1}{2^{2K}} \binom{2K}{K}. \tag{22}$$

E.2 S_0 **for** $p = 0.5$, $a_G = a_R = 1$ **and** $\theta = 0.5$ In this case the threshold is a noninteger value, $\theta = 0.5$. Therefore, from Eq. (9) and (10), we have

$$S_0 = K I^{(1)}. \tag{23}$$

where from Eq. (5c)

$$I^{(1)} = \frac{1}{2^{2K}} \sum_{i=0}^{K} \binom{K}{i} \sum_{l=0}^{i} \sum_{m=0}^{K-i} \binom{i}{l}\binom{K-i}{m} \mathcal{I}(a_G = 1, a_R = 1, \theta = 0.5, b = 0.5). \tag{24}$$

In order to calculate $\mathcal{I}(a_G = 1, a_R = 1, \theta = 0.5, b = 0.5)$ we consider the following. The noninteger threshold $\theta = 0.5$ makes all the terms with Kronecker deltas effectively zeros, since the exact value of the threshold cannot be attained. Given that $\theta = 0.5 > 0$, it also makes nodes in which the sum of the updating function equals 0, $\sum_j a_{n,j}\sigma_{n_j} = 0$, become inactive, as this sum is smaller than the threshold. As a consequence, the only way in which the changing of one of the inputs of a node changes the state of the target node is if either the sum before the damage was 0 and after the damage it is above 0, or vice versa. In addition, since $k_d = 1$, then the only possible values for u, v, w and $z = k_d - u - v - w$ are 1 on one of them and 0 on the rest. Using these facts and Eq (5) we get

$$\mathcal{I} = \sum_{u=u_0}^{u_f} \sum_{v=v_0}^{v_f} \sum_{w=w_0}^{w_f} Pr(u, v, w) \{H(l - u + w - m + v - z - 0.5)$$

$$\times H(m - l + 0.5) + H(m - v + z - l + u - w + 0.5)H(l - m - 0.5)\}$$

$$= \left[\frac{m}{K} + \frac{i-l}{K}\right] \delta_{l,m} + \left[\frac{l}{K} + \frac{K-i-m}{K}\right] \delta_{l,m+1}$$

$$= \frac{K-i+1}{K} \delta_{l,m+1} + \frac{i}{K} \delta_{l,m}. \tag{25}$$

Using this result in Eq. (24) we have

$$I^{(1)} = \frac{1}{2^{2K}} (g_1 + g_2), \tag{26}$$

where

$$g_1 = \frac{1}{K} \sum_{i=0}^{K} \binom{K}{i} \sum_{l=0}^{i} \sum_{m=0}^{K-i} \binom{i}{l}\binom{K-i}{m} \delta_{l-1,m}(K-i+1), \qquad (27)$$

$$g_2 = \frac{1}{K} \sum_{i=0}^{K} \binom{K}{i} \sum_{l=0}^{i} \sum_{m=0}^{K-i} \binom{i}{l}\binom{K-i}{m} i\delta_{l,m}. \qquad (28)$$

From Eqs. (13) and (27) it follows that

$$g_1 = 2f_1 = \frac{1}{2}\binom{2K}{K}, \qquad (29)$$

Thus, we only need to calculate g_2. By using Vandermonde's identity and the mean value of the hypergeometric function, Eqs. (16) and (17), we obtain

$$\begin{aligned}g_2 &= \frac{1}{K} \sum_{i=0}^{K} \binom{K}{i} i \sum_{l=0}^{i} \binom{i}{l}\binom{K-i}{l} \\ &= \frac{1}{K} \sum_{i=0}^{K} \binom{K}{i} i \sum_{l=0}^{i} \binom{i}{i-l}\binom{K-i}{l} \\ &= \frac{1}{K} \sum_{i=0}^{K} \binom{K}{i}\binom{K}{i} i \\ &= \frac{1}{K} \sum_{i=0}^{K} \binom{K}{i}\binom{K}{K-i} i = \frac{1}{K}\frac{K}{2}\binom{2K}{K} = \frac{1}{2}\binom{2K}{K}. \qquad (30)\end{aligned}$$

Using Eqs. (26), (29) and (30) in Eq. (23), we get the desired result (the $\theta = 0.5$ case of Eq. (12)):

$$S_0\,[p=0.5, a_G=1, a_R=-1, \theta=0.5] = \frac{K}{2^{2K}}\binom{2K}{K}. \qquad (31)$$

E.3 $S_0\,(p=0.5, a_G=1, a_R=1, \theta=-0.5)$ Since, $\theta = -0.5$, from Eqs. (9), (10) and (5c) we get

$$S_0 = K I^{(1)}, \qquad (32)$$

$$I^{(1)} = \frac{1}{2^{2K}} \sum_{i=0}^{K} \binom{K}{i} \sum_{l=0}^{i} \sum_{m=0}^{K-i} \binom{i}{l}\binom{K-i}{m} \mathcal{I}(a_G=1, a_R=1, \theta=-0.5, b=0.5). \qquad (33)$$

To calculate $\mathcal{I}(a_G = 1, a_R = 1, \theta = -0.5, b = 0.5)$ we look back at the derivation of $\mathcal{I}(a_G = 1, a_R = 1, \theta = 0, b = 0.5)$ in the last section (Sec. 5) and notice that both cases are similar. The difference lies in that, in this case, the threshold $\theta = -0.5 < 0$. This threshold then makes the nodes in which the sum of the updating function equals 0, (namely for which $\sum_j a_{n,j}\sigma_{n_j} = 0$), become active, as the sum is larger than the threshold. Therefore, the only cases in which damaging one of the inputs of a node changes the state of the target node is if either the sum before the damage was 0 and after damage it is below 0, or vice versa. Again, since $k_d = 1$, the only possible values for u, v, w and $z = k_d - u - v - w$ are 1 on one of them and 0 on the rest. Using all this in Eq. (5) we obtain

$$\mathcal{I} = \sum_{u=u_0}^{u_f} \sum_{v=v_0}^{v_f} \sum_{w=w_0}^{w_f} Pr(u,v,w) \{H(l - u + w - m + v - z + 0.5)$$
$$\times H(m - l - 0.5) + H(m - v + z - l + u - w - 0.5)H(l - m + 0.5)\}$$
$$= \left[\frac{m}{K} + \frac{i-l}{K}\right]\delta_{l+1,m} + \left[\frac{l}{K} + \frac{K-i-m}{K}\right]\delta_{l,m}$$
$$= \frac{i+1}{K}\delta_{l+1,m} + \delta_{l,m} - \frac{i}{K}\delta_{l,m}. \tag{34}$$

Using this result in Eq. (24) we get

$$I^{(1)} = \frac{1}{2^{2K}}(h_1 + h_2 - h_3), \tag{35}$$

where

$$h_1 = \frac{1}{K}\sum_{i=0}^{K}\binom{K}{i}\sum_{l=0}^{i}\sum_{m=0}^{K-i}\binom{i}{l}\binom{K-i}{m}\delta_{l+1,m}(i+1), \tag{36}$$

$$h_2 = \sum_{i=0}^{K}\binom{K}{i}\sum_{l=0}^{i}\sum_{m=0}^{K-i}\binom{i}{l}\binom{K-i}{m}\delta_{l,m}. \tag{37}$$

$$h_3 = \frac{1}{K}\sum_{i=0}^{K}\binom{K}{i}\sum_{l=0}^{i}\sum_{m=0}^{K-i}\binom{i}{l}\binom{K-i}{m}i\delta_{l,m}. \tag{38}$$

By comparing Eq. (14) with (36) and Eq. (28) with (38), we have

$$h_1 = 2f_2 = \frac{1}{2}\binom{2K}{K}, \tag{39}$$

$$h_3 = g_2 = \frac{1}{2}\binom{2K}{K}. \tag{40}$$

The only term that has not been calculated yet is h_2. To do this we use Vandermonde's identity, Eq. (16), with $m = n = K$

$$\begin{aligned}
h_2 &= \sum_{i=0}^{K} \binom{K}{i} \sum_{l=0}^{i} \binom{i}{l}\binom{K-i}{l} \\
&= \sum_{i=0}^{K} \binom{K}{i} \sum_{l=0}^{i} \binom{i}{i-l}\binom{K-i}{l} \\
&= \sum_{i=0}^{K} \binom{K}{i}\binom{K}{i} \\
&= \binom{2K}{K}.
\end{aligned} \qquad (41)$$

Finally we get the $\theta = -0.5$ case of Eq. (12) using Eqs. (39), (40), (41) and (35) in Eq. (32),

$$S_0\left(p = 0.5, a_G = 1, a_R = 1, \theta = -0.5\right) = \frac{K}{2^{2K}}\binom{2K}{K}, \qquad (42)$$

which is, of course, the same as the case $\theta = 0.5$ because of the symmetry of the weights and the probabilities.

Structure-Dynamics Relationships in Biological Networks

Juha Kesseli[1], Lauri Hahne[1], Olli Yli-Harja[1],
Ilya Shmulevich[2], and Matti Nykter[1]

[1] Department of Signal Processing, Tampere University of Technology,
[2] Institute for Systems Biology, Seattle, USA

Abstract. Biological systems at all scales of organization display exceptional abilities to coordinate complex behaviors to balance stability and adaptability in a variable environment and to process information for mounting diverse, yet specific responses. Such emergent macroscopic behaviors are governed by complex systems of interactions. Understanding how the structure of such interactions determines global dynamics is a major challenge in complex systems theory. Traditionally this problem has been approached by defining statistics that quantify some aspect of structure or dynamics of complex systems. Such an analysis has helped to gain insight into the manner how particular structural properties such as topology constrains dynamical behavior. However, a drawback of this analysis is that established relationships between structural properties and dynamical bahaviour are specific to the measure, and thus hold only on this single aspect. Thus, it is difficult to establish more general principles. To overcome these limitations we propose to use information theory to establish a unified framework through which one can begin to examine both structure and dynamics of complex systems in a general manner.

1 Introduction

In biology it becomes quickly apparent that knowing the parts and the manner in which they interact is often insufficient for understanding the global emergent behavior of the system. This is of course true even of physical systems. Already in the 19th century, it was recognized that knowing the interactions between two bodies was not enough to understand or solve completely the dynamics of a group of such bodies, even in the context of classical Newtonian laws of motion and gravity [1].

It is thus a paramount challenge to understand how the structure or topology of a complex system constrains and determines the repertoire of the global dynamical processes available to the system. However, it is important to be able to understand such relationships not in terms of specific physical processes or reactions between chemical reactants, but instead in terms of entities or agents that interact via certain rules, so that what is learned can be generalized to many systems sharing common characteristics at different scales of organization.

In biological systems, the structure of the interactions determines emergent dynamical behaviors and properties, such as robustness and homeostasis, multi-stability, adaptability, decision-making, and information processing [2,3]. However, understanding the relationships between topology and dynamics in complex systems is a fundamental but widely unexplored problem [4,5]. Perhaps one reason why general principles behind such relationships are still lacking is, in part, due to the lack of sufficiently general formalisms for representing and analyzing the information embedded in the structures and dynamics of complex systems within a common theoretical framework that is not implicitly tied to particular model classes or features.

Numerous relationships between specific structural and dynamical features of networks have been investigated [2,6,7]. For example, structure can be studied by means of various graph theoretic features of network topologies such as degree distributions [8] or modularity [9], or in terms of classes of updating rules that generate the system dynamics [10]. Aspects of dynamical behavior include transient and steady-state dynamics as well as responses of a system to perturbations [11].

Information theory provides a common lens through which one can study both the structure and the dynamics of complex systems within a unified framework. Indeed, since network structures as well as their dynamic state trajectories are objects that can be represented on a computer, the information encoded in both can be compared and related using appropriate information theoretic tools. Unlike Shannon's information, which pertains to distributions of objects [12], Kolmogorov complexity is a suitable framework for capturing the information embedded in individual objects of finite length as well as the information shared between objects [13].

In this chapter, we will review various measures for structure and dynamics of complex networks. We will show how these measures can be used to quantify aspects of dynamical behavior of the systems using Boolean networks as a model system. To go beyond the summary statistics of networks, we propose to use information theory as a unified approach to establish a link between structural properties and dynamical behavior. To this end, recent developments in information theory are discussed and applicability of the proposed approach is demonstrated.

The structure of this chapter is as follows. First, we present some background on the models and methodologies that are used in this work. Then we will review various measures for quantifying structure and dynamics of complex systems. We will demonstrate how these measures can be used to study structure dynamics relationships. Finally, we will use information theory to established a direct link between structure and dynamics. Applications of recent information theoretical tools will be demonstrated.

2 Background

2.1 Boolean Networks as Models of Complex Dynamical Systems

A Boolean network (BN) is a directed graph with N nodes. Nodes represent elements of the system and graph arcs represent interactions between the elements.

Each node is assigned a binary output value and a Boolean function, whose inputs are defined by the graph connections. Let $s_i(t) \in \{0,1\}$, $i = 1, \ldots, N$, where N is the number of nodes in the network, be the state of i:th node in a Boolean network at time t. The state of this node at time $t+1$ is determined by the states of nodes $j_1, j_2, \ldots, j_{k_i}$ at time t as

$$s_i(t+1) = f_i(s_{j_1}(t), s_{j_2}(t), \ldots, s_{j_{k_i}}(t)), \qquad (1)$$

where $f_i : \{0,1\}^{k_i} \to \{0,1\}$ is a Boolean function of k_i variables. A binary vector $\boldsymbol{s}(t) = (s_1(t), \ldots, s_N(t))$ is the state of the network at time t. In a synchronous BN all nodes are updated simultaneously as the system transitions from state $\boldsymbol{s}(t)$ to state $\boldsymbol{s}(t+1)$ [11].

Random Boolean networks (RBNs) are networks in which each node has exactly K inputs that are selected randomly. The update rules are chosen with bias b, such that for an update rule f $E[f(x)] = b$ for any input x and for any $x \neq y$ $f(x)$ and $f(y)$ are selected independently. In addition to this narrow sense of the word, random Boolean networks can also be used to describe networks generated with some other selected distribution or pattern of update rules, for example. These assumptions of randomness permit analytical insights of the behavior of large networks. RBNs were used as the first model of GRN [14]. Each node is a gene, and is assigned a Boolean function from the set of possible Boolean functions of k variables.

By running the network over several time steps starting from an initial state, a trajectory through the network's state space can be observed (referred to as a "time series"). Over time, the system follows a trajectory that ends on a state cycle attractor. In general, a RBN has many such attractors. It should be noted that this model can directly be generalized to a larger alphabet by defining $s_i(t) \in \{0, \ldots, L-1\}$ and $f_i : \{0, \ldots, L-1\}^{k_i} \to \{0, \ldots, L-1\}$, where L is the size of the alphabet.

Dynamical regime of Boolean networks. One important feature of RBNs is that their dynamics can be classified as ordered, critical, or chaotic. During the simulation of a RBN some nodes will become "frozen", meaning that they will no longer change their state, while other will remain dynamic, meaning that there state will periodically change from one state to the other. The fractions of frozen and dynamic nodes depends on the network dynamical regime.

In "ordered" RBNs, the fraction of nodes that remain dynamical after a transient period vanishes like $1/N$ as the system size N goes to infinity; almost all of the nodes become "frozen" on an output value (0 or 1) that does not depend on the initial state of the network. In this regime the system is highly stable against transient perturbations of individual nodes, meaning that externally imposing a change in one node state will not cause significant changes in the other nodes states. In "chaotic" RBNs, the number of dynamical, or "unfrozen" nodes scales like N and the system is unstable to many transient perturbations, meaning that a perturbation will spread through many nodes.

Here we consider ensembles of RBNs parameterized by the average indegree K (i.e, average number of inputs to the nodes in the network), and the bias p (i.e., the fraction of inputs states that lead to an output with value "1") in the choice of Boolean rules. The indegree distribution is Poissonian with mean K and at each node the rule is constructed by assigning the output for each possible set of input values to be 1 with probability p, with each set treated independently. If $p = 0.5$, the rule distribution is said to be unbiased. For a given bias, the critical connectivity, K_c, is equal to [15]:

$$K_c = [2p(1-p)]^{-1}. \quad (2)$$

For $K < K_c$ the ensemble of RBNs is in the ordered regime; for $K > K_c$, the chaotic regime. For $K = K_c$, the ensemble exhibits critical scaling of the number of unfrozen nodes; e.g., the number of unfrozen nodes scales like $N^{2/3}$. The order-chaos transition in RBNs has been characterized by several quantities. These will be reviewed in section 3.2.

2.2 Measures of Information

Here some fundamental results of information theory and interesting new developments are discussed. The presented results will form basic tools that allow us to study structure and dynamics under unified framework.

There are two commonly used definitions for information, Shannon information [16] and Kolmogorov complexity [17,18,19]. Both theories provide a measure of information using the same unit: a bit. A natural interpretation of information is the length of the description of an object in bits.

Information distance. In Shannon information theory the amount of information is measured by entropy. For a discrete random event x with k possible outcomes, the entropy H is given as

$$H = \sum_{i=1}^{k} p_i I_i = -\sum_{i=1}^{k} p_i \log p_i, \quad (3)$$

where p_i is the probability of an event x_i to occur [20]. Quantity $I_i = -\log p_i$ is the information content of an event x_i. Natural interpretation for entropy is that it is the expected number of bits that are needed to encode the outcomes of a random event x. It can be observed that entropy is maximized when the probabilities of all events are equal, that is $p_i = \hat{p}, \forall i \in 1, \ldots, k$ [20].

Shannon information measures information of a distribution. Thus, it is based on the underlying distribution of the observed random variable realizations. The distribution can be obtained based on assumptions about the data generation process or it can be estimated from the data. Thus, to utilize Shannon information, the alphabet of the data source needs to be fixed and there needs to be a model for the origin of the data.

While Shannon information has successfully been used for the analysis of complex systems, for example by studying the mutual information in dynamics or the networks ability to store information [21,22], when studying structure and dynamics of complex systems, it is not clear how to define a unified alphabet and the underlying distribution. Thus, for these analyses the alternative definition of information, Kolmogorov complexity, seems a more attractive option.

Unlike Shannon information, Kolmogorov complexity or algorithmic information is not based on statistical properties, but on the information content of the object itself [13]. Thus, Kolmogorov complexity does not consider the origin of an object. The Kolmogorov complexity $K(x)$ of a finite object x is defined as the length of the shortest binary program that with no input outputs x on a universal computer. Thus, it is the minimum amount of information that is needed to generate x. Unfortunately, in practice this quantity is not computable [13].

While the computation of Kolmogorov complexity is not possible, an upper bound can be estimated using lossless compression [13]. Several real-life compression algorithms, like the Huffman [23], Lempel-Ziv [24], and arithmetic coding [25] have proven to give useful approximations of Kolmogorov complexity in practical applications [13].

As information is an absolute measure, related to a single object or a distribution, it is not directly suitable for comparing the similarities of two objects. Small or large information alone does not tell much about the similarity of objects. Thus, measures to jointly compare the information content of two objects have been proposed.

Information-based similarity measures can be defined in terms of Kolmogorov complexity. This topic has been studied in recent years with the goal of finding an information measure than can be approximated computationally [26,27].

We denote as $K(x, y)$ the length of the shortest binary program that outputs x and y, and a description how to tell them apart. We can define a conditional Kolmogorov complexity $K(x|y)$ as the length of the shortest binary program that with a given input y outputs x [13]. Thus, information about y, contained in x can be defined as [27]

$$I(x; y) = K(y) - K(y|x). \qquad (4)$$

It can be shown that the relation

$$K(x, y) = K(x) + K(y|x) = K(y) + K(x|y) \qquad (5)$$

holds up to an additive precision [13]. Therefore, there exists a symmetry property $I(y; x) = I(x; y)$, up to an additive precision.

Kolmogorov complexity based similarity measure, or information distance, between two objects is the shortest binary program that computes x from y, or vice versa. Thus, information distance can be defined as [26]

$$d_{ID}(x, y) = \max(K(y|x), K(x|y)). \qquad (6)$$

This is a measure of absolute information distance between two objects. As the size of an object has a direct impact to the Kolmogorov complexity of the object, we should define a normalized version of the information distance that takes the size of an object into account. A normalized information distance can be defined as [27]

$$d_{NID}(x,y) = \frac{\max(K(x|y), K(y|x))}{\max(K(x), K(y))}. \quad (7)$$

While normalized information distance can be motivated solely from the information theory point of view, it has some general properties that make it interesting in other ways. The normalized information distance has been shown to incorporate all effective computable distance metrics including, for example, the Euclidean and Hamming distances. Thus, the normalized information distance can be argued to be a universal measure of similarity.

Normalized Compression Distance. While normalized information distance, like Kolmogorov complexity itself, is not computable, it has been shown that this metric can be approximated by any real-life compression algorithm that fulfills several natural criteria of a *normal compressor* C, see [28] for details.

By using a compressor C instead of the Kolmogorov complexity K, we can write Equation 7 in a computable form. After we apply Equation 5, to the the numerator of Equation 7, the numerator can be written as [27]

$$\max\{K(x,y) - K(y), K(x,y) - K(x)\}.$$

For compression convenience we can approximate $K(x,y)$ by the concatenation of these strings: $K(x,y) = K(xy) = K(yx)$ holds upto an additive precision. Using these properties the normalized compression distance (NCD) can be defined as

$$d_{NCD}(x,y) = \frac{C(xy) - \min(C(x), C(y))}{\max(C(x), C(y))}. \quad (8)$$

It can be shown that this approximation has the same metric properties as the normalized information distance, up to an additive constant [27,28].

Set complexity. As discussed above, information distance captures the similarity between two objects. These objects could be the structure and the dynamics of a given network, or pairs of structures and dynamics from an ensembles of networks. However, to gain more insight into structure dynamics relationships, a more general measure that captures relationships between ensembles of objects or features such as the nodes of the network would be useful. For this purpose, a measure called set complexity was recently proposed.

Instead of just a pairwise similarities, Set complexity quantifies the information stored in a set S of strings x_i. This set can be constructed from any strings or objects that can be represented on computer. Key property of this measure is that it reflects the information stored in individual strings as well as the information stored in the set itself. More importantly, the measure is effective in the sense that it has the properties: 1) a random string adds zero information to the set, 2) Duplicated strings add little or no information to the set.

Set complexity, that fulfills these conditions, is defined as

$$\Psi(S) = \sum_{i=1}^{N} K(x_i) F_i(S), \tag{9}$$

where

$$F_i(S) = \frac{2}{N(N-1)} \sum_{pairs\ i,j} d_{ij}(1 - d_{ij}) \tag{10}$$

and d_{ij} is the normalized information distance between x_i and x_j. This satisfies the properties defined earlier because for completely random strings $1-d_{ij}$ should be zero if the strings are long enough and for identical strings d_{ij} is zero. In addition, the measure clearly takes into account both the individual information of the strings and their relation to each other. It has been shown earlier that just as with NCD, this measure of set complexity can also be approximated by using real life compression algorithms to estimate uncomputable information distance.

3 Summary Statistics for Dynamical Systems

3.1 Measures of Structural Properties

Properties of a network are related to its structure. Traditionally, networks have been analyzed assuming that the connections between different nodes are selected randomly [29,14,11]. Recent discoveries have shown that this assumption does not hold for most real world networks [30,31,32,33]. Instead, several networks, including gene regulatory networks show a scale free type of structure [31]. A characteristic property of a scale free network is the existence of hubs, that is, the nodes that have a very high number of connections. In a random topology all the nodes have approximately the same number of connections.

Structure of the network can be characterized using summary statistics, that can be computed for any given network [9,34]. To quantify the topology, most direct measures are the in- and outdegree of the network. For a directed graph, indegree and outdegree measure the connectedness of individual nodes. Indegree measures the incoming edges of a node and outdegree the outgoing edges. For a given graph $G = (V, E)$, V is the set of nodes and E is the set of edges, applies

$$\sum_{v \in V} \deg^+(v) = \sum_{v \in V} \deg^-(v) = |E|, \tag{11}$$

where $\deg^+(v)$ is the out and $\deg^-(v)$ is the in degree of node v.

Clustering coefficient is a measure for the connectivity of a network [9]. It is defined for a given node as the number of neighboring nodes that are connected to each other. That is, for a set of nodes $N = n_1, \ldots, n_k$ we have a set of connections (edges) $E = \{e_{ij}\}$, where $i, j \in 1, \ldots, k$. Thus e_{ij} is an edge between the nodes n_i and n_j. We can define a neighborhood B for the node n_i as its immediately connected neighbors $B_i = \{n_j\} : e_{ij} \in E$. The connectivity k_i of the node n_i is the size of the neighborhood $|B_i|$. The clustering coefficient C_i for the node n_i is the proportion of links between the neighborhood nodes divided by the number of links that could possibly exist. For each neighborhood the maximum number of links is $k_i(k_i - 1)$. Thus, the clustering coefficient is given as

$$C_i = \frac{|\{e_{lm}\}|}{k_i(k_i - 1)} : n_l, n_m \in B_i, e_{lm} \in E. \qquad (12)$$

The clustering coefficient for the whole network is the average of the clustering coefficients of all the nodes

$$\hat{C}_i = \frac{1}{n} \sum_{i=1}^{k} C_i. \qquad (13)$$

Another measure of network topology is characteristic path length [9]. First, the path length L_{ij}, that is the minimum number of edges that are needed to get from the node n_i to the node n_j, is computed. The characteristic path length L is then L_{ij} averaged over all pairs of nodes.

3.2 Measures of Dynamical Behavior

In this section, we present the basics of annealed approximation, a technique utilized to characterize the dynamics of Boolean networks. This is relevant in the context of studying structure-dynamics relationships, since it provides a simple analysis in the case of networks with no local structure. Any effects caused by the topology of the network, causing correlated values in the nodes of the network, can then be seen as deviations of this baseline dynamical network characterized by average sensitivity, presented below. We also introduce a numerical measure that can be computed by simulating networks, and which generalizes the average sensitivity so that it can also capture a part of the topological effects on dynamics missed by the standard analysis of annealed approximation. In addition, we present the concept of the frozen core, which is used to characterize the network dynamics in an alternative way, taking more fully into account the steady state behavior of the dynamics caused by the topology.

Standard analysis. Annealed approximation in Boolean networks was presented as a way to give an analytical derivation for the numerically observed differences in behavior between ordered, critical and chaotic random Boolean networks [15,35]. The approach is probabilistic: expected short-term behavior of networks over a distribution of characteristic states is calculated and this expected value is used to predict the expected long-term behavior of a network taken from the distribution. The distribution of characteristic states can be selected or parametrized in different ways, and depending on this selection different approximations are obtained. For the purposes of this chapter, the state of the network will be described with two values, the state bias b_t (the probability that a node obtains value 1 at time t) and the probability that a node is perturbed, ρ_t. It is, further, assumed that these two properties are independent. This holds in many simple cases, including when the functions of the network are selected to have random functions with some bias p. In addition, in this particular case, b_t remains constant.

By an annealed network we mean a Boolean network in which the connections in the network are reshuffled after each update. This has the effect of, first of all, breaking up the attractor structure. Secondly, the network nodes in general lose their identity in the sense that two states that are identical up to the ordering of the nodes are, in effect, the same state in an annealed network. The quenched network, in which this annealing is not performed, can have local topological structures that affect the network dynamics. Using annealed approximation means that we take the results computed in the annealed model and apply them to quenched random networks. For the purposes of the annealed approximation all we have left of the topological properties is contained in the distribution of in-degrees of the nodes in the network.

In this case, the distribution of characteristic states, i.e. states identical up to a permutation of nodes, can successfully be parametrized by a single value ρ_t. This value is the probability that an arbitrary node is perturbed at time t. The nodes are assumed to be independent. As a result, it was found that the parameter values at the phase transition are given by setting what is now called the average sensitivity to one, $2Kp(1-p) = 1$. If the average sensitivity is less than one, perturbations are predicted to die out on average and the network is called ordered. If $2Kp(1-p) > 1$, the network is chaotic and perturbation size will approach a non-zero fixed point.

If this case is extended to cover an arbitrary distribution $\mathfrak{F}$ of functions in the network the change in perturbation size ρ from time t to time $t+1$ can be described by the iterative map h_1, $\rho_{t+1} = h_1(\rho_t)$ where

$$h_1(\rho_t) = \mathop{E}_{f \in \mathfrak{F}} \left[\frac{1}{2^{K_f}} \sum_{x \in \mathbb{B}^{K_f}} \sum_{y \in \mathbb{B}^{K_f}} f(x) \oplus f(y) (1-\rho_t)^{K_f - |x \oplus y|} (\rho_t)^{|x \oplus y|} \right]. \quad (14)$$

The fixed point ρ^*, $\rho^* = h_1(\rho^*)$ of this mapping is used to predict the chaoticity of the network. If the fixed point is non-zero, $\rho^* \neq 0$, this means that small enough perturbations will grow on average towards the stable-state value and the network is called chaotic. From the map, this property may be determined by computing $h'_1(0)$, which is, in fact, a way of computing the average sensitivity for a general distribution of functions.

Chaotic networks correspond to the case $h'_1(0) > 1$. If the network has a fixed point at the origin, $\rho^* = 0$, all perturbations will eventually die out according to this annealed approximation. If $h'_1(0) < 1$ the network is called ordered and as a limiting case if $h'_1(0) = 1$ the network is called critical. h_1 is commonly called the Derrida map or the Derrida curve of the network, and numerical approximations in particular can be called Derrida plots. In the special case of random Boolean networks in the narrow sense (with random connectivity), this approximation is sufficient for determining the chaoticity of quenched networks with selected p and K. In the following, an alternative form of the analysis is presented that enables the study of a wider class of Boolean networks. Similar classification of networks based on the fixed points of perturbation size can be made in the following case as well.

Generalized average sensitivity. In [36] a general form describing the annealed model is given in the form of four iterative maps. In this mapping, the perturbation is parametrized with probabilities p_{ij} that describe the probability that the state of an arbitrary node has value i without a perturbation and value j with perturbation applied to the network. This iterative mapping is in effect three-dimensional since the sum of p_{ij} is equal to one. The characteristic states of the network can thus in this approximation be described with three parameters. Different kinds of Derrida maps are explicitly derived from this general framework in [36]. The interest in Derrida maps is justified despite the three-dimensional nature of the mentioned annealed approximation, since one-dimensional mappings are simpler to analyze in terms of their fixed point behavior. In addition, applying the three-dimensional map to perturbation spreading gives one-dimensional mappings in a natural way.

In addition to the Derrida maps describing perturbation propagation, an additional iterative map called the bias map is used to describe the evolution of the proportion of ones in the state of a network. With the help of the bias map, versions of the Derrida map can be used to capture fixed-point behavior of the annealed model. The bias map can be written as

$$g(b) = \underset{f \in \mathfrak{F}}{E}\left[\sum_{x \in \mathbb{B}^{K_f}} f(x) P(x|b) \right],$$

where

$$P(x|b) = b^{|x|}(1-b)^{K_f - |x|}$$

is the probability for input vector x given that we know probability b for an input to have value 1. This mapping is contained in the iterative maps for p_{ij} and can be obtained from the update equation for p_{11} by setting $p_{01} = p_{10} = 0$.

The bias map can be iterated by

$$b_{t+1} = g(b_t).$$

This mapping may have non-trivial fixed point solutions depending on the chosen function distribution [37,38,39]. In [40] a definition is given for stable functions as ones that have a bias map fixed point at zero or one. If that occurs, the network constructed of these functions will necessarily be stable without further study of Derrida maps being needed. In annealed models used for biological applications we typically assume for modeling purposes that the bias will reach some fixed point $b^* = g(b^*)$. If we study e.g. perturbation propagation we can then use this fixed point of the annealed model to correspond to states on the attractor in the quenched model. In this chapter, we quantify network dynamics in the fixed point of the state bias.

For a function distribution $\mathfrak{F}$ average influence $I(b)$ is

$$I(b) = \underset{f \in \mathfrak{F}}{E}\left[\frac{1}{K_f}\sum_{i=1}^{K_f}\sum_{x \in \mathbb{B}^{K_f}} f(x) \oplus f(x \oplus e_i)P(x|b)\right].$$

We define average influence I of function distribution $\mathfrak{F}$ at bias-map fixed point b^* as

$$I = I(b^*).$$

Average influence I can be considered as the average probability that an arbitrary arc is propagating a perturbation at the fixed point state. Influence of variable i is defined as

$$I_i(b) = \sum_{x \in \mathbb{B}^{K_f}} f(x) \oplus f(x \oplus e_i)P(x|b).$$

The average sensitivity of a Boolean function is the sum of influences $I_i(b)$. The average sensitivity of a function distribution is given by

$$\lambda(b) = \underset{f \in \mathfrak{F}}{E}\left[\sum_{i=1}^{K_f}\sum_{x \in \mathbb{B}^{K_f}} f(x) \oplus f(x \oplus e_i)P(x|b)\right].$$

By using the average sensitivity at the bias map fixed point we can define the network's average sensitivity as

$$\lambda = \lambda(b^*).$$

$\lambda(\frac{1}{2})$ has also been used for the purpose, but this can be misleading in cases in which b^* differs significantly from $\frac{1}{2}$ [41].

λ is the average amount of nodes that are perturbed one time step after we have flipped the value of a randomly chosen node, given that the network has reached the bias map fixed point before the perturbation. In its asymptotical nature $\ln \lambda$ can be considered to correspond to the Lyapunov exponent in the classical theory of chaotic systems, although the analogy should not be streched too far.

In terms of Derrida maps, we can select the characteristic states so that one state is characterized by its bias b_1 alone and the second state is characterized by the probability ρ of any one of its bits being different from the first state. In this case the Derrida map is derived from the general iterative maps in [36] by setting $b_1 = b^*$ and $b_2 = b^*(1-\rho) + \rho(1-b^*)$. $h_2(\rho)$ is obtained as

$$h_2(\rho) = \underset{f \in \mathfrak{F}}{E}\left[\sum_{x \in \mathbb{B}^{K_f}}\sum_{y \in \mathbb{B}^{K_f}} f(x) \oplus f(y)\left((1-b^*)(1-\rho)\right)^{(1-x)^T(1-y)}\cdots\right.$$

$$\left. \left(b^*(1-\rho)\right)^{x^T y}\left(\rho(1-b^*)\right)^{(1-x)^T y}\left(b^*\rho\right)^{x^T(1-y)}\right].$$

This expression can be simplified as

$$h_2(\rho) = \underset{f \in \mathfrak{F}}{E}\left[\sum_{k=1}^{K_f} \lambda_k \rho^k (1-\rho)^{K_f - k}\right], \qquad (15)$$

where

$$\lambda_k = \sum_{x \in \mathbb{B}^{K_f}} \sum_{y \in P_k} f(x) \oplus f(x \oplus y) P(x|b^*)$$

is the average sensitivity of function f over k variables. λ_k is also called the generalized sensitivity in [42]. The slope of this Derrida curve corresponds to our utilization of generalized sensitivity $\lambda(b^*)$ above.

Since we lose information when we use one-dimensional Derrida maps instead of the three-dimensional map from which these maps can be obtained, we should in general use the original map to study the behavior of the system. In any case, the annealed approximation in this form is limited in its ability to represent information propagation in networks due to the limitation to average quantities instead of distributions. A crude approximation of the dynamics may, however, be obtained in this form.

Quantifying dynamics with correlations. If the network had an infinite size and did not have local topological structure so that its dynamics can really be accurately approximated with the annealed approximation presented above, the states of the network would be well described with the independent node model. In practice, finite size and topological structures cause the dynamics to deviate from this simple baseline dynamics. In order to quantify the dynamics in this, more realistic case, we compute parts of Derrida plots numerically over several time steps. In the case of independent nodes, the results could be predicted by iterating the Derrida map, with each iteration obtaining a prediction one time step further. Difference from the numerical Derrida map over several time steps therefore quantifies the effects of topology.

In general, we can select a few time lags $m_1, m_2, ..., m_M$ for which we compute the perturbation size. Starting from networks that we run for long enough for the initial transient to settle, we toggle a single bit in the state and follow the trajectories of both the perturbed and the unperturbed network. After m_i timesteps we record the average number of nodes perturbed from a large number of individual trials. This average can be denoted by $\hat{\lambda}^{(m_i)}$, and the values thus obtained give a simple description of the network dynamics over the range of time steps studied. These values give, in fact, estimates of the generalized average sensitivity over m_i time steps. For the purpose of illustrations in this chapter, we select $M = 1$ and $m_1 = 5$.

Frozen core. Another measure of dynamics considered in this chapter is the size of the frozen core (see section 2.1). A frozen core arises in networks due to, for example, constant functions, the outputs of which will always have the same value after an update. These nodes are called frozen. Further, there are some nodes which obtain their inputs only from frozen nodes or from nodes whose values always have a combination that will cause the output to obtain the same value. These nodes, obtaining the same constant value for all the attractor states, are included in the frozen core as well. The total size of the frozen core is selected as a dynamical parameter we quantify from all the networks of the chapter. For ordered networks, most of the network is frozen, whereas for chaotic networks, the frozen core is tiny.

4 Structure-Dynamics Relationships

4.1 Relationships in Terms of Summary Statistics

Topological statistics like clustering coefficient and characteristic path length can be used to determine the type of network and to compare different network topologies [9]. For a network with regular wiring the clustering coefficient and characteristic path length are both high, whereas in a random network both statistics have a small value. Other network topologies can have a high clustering coefficient and still a low average path length. Networks with this kind of topology are known as small world networks and they have the property that $L > L_r$ but $C \gg C_r$, where L_r and C_r denotes the characteristic path length and clustering coefficient of a random network, respectively. Usually this kind of a network also has a scale free topology [9].

While these topological statistics can successfully be used to compare and classify different types of networks, it is not obvious what measures are able the uncover all the interesting characteristics of a network. Furthermore, measures like the characteristic path length and clustering coefficient are most useful in the comparison of different topologies. They are not that informative when, for example, two scale free networks are compared. This is a problem if we want to compare networks that have the same topological properties.

To show how different network summary statistics can be used to study structure-dynamics relationships, we generated networks from six ensembles. Three of the ensembles are the standard random Boolean networks with in-degrees $K = 1$, $K = 2$ and $K = 3$ corresponding to ordered, critical and chaotic dynamics, respectively. The three others are built by forming a lattice with a similarly varying in-degree and changing 5% of the inputs but keeping the in-degree of the nodes the same. Our test cases correspond, therefore, to a topologically homogeneous network with no local structure and a network with a high degree of regular local structure. Distributions of different structure and dynamics parameters for these ensembles are illustrated in Figs. 1 – 4.

To understand the relationships in terms of these parameters we can link them together using straightforward correlation analysis. Scatter plots in Figs. 5–8 show that using different measures of structure and dynamics different features of the relationships can be seen.

These illustrations show that the view on the structure dynamics relationships is very different depending on the parameters we look at. It is worth noting that none of the studied pairs of parameters are able to make all the network ensembles distinguishable from each other. In addition, while the distributions

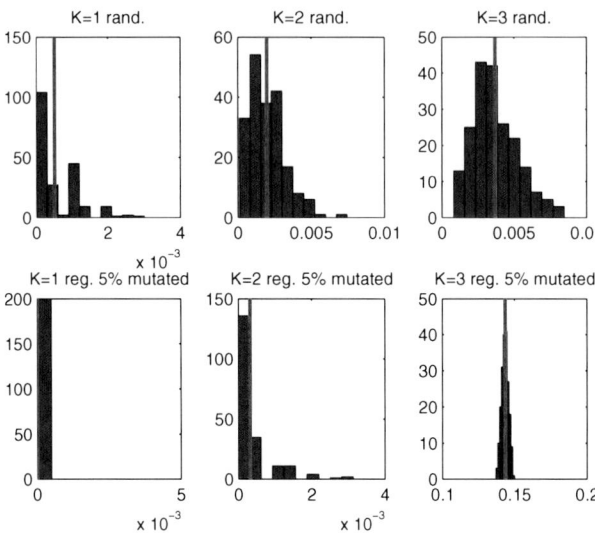

Fig. 1. Distributions of the clustering coefficients of networks from different ensembles.

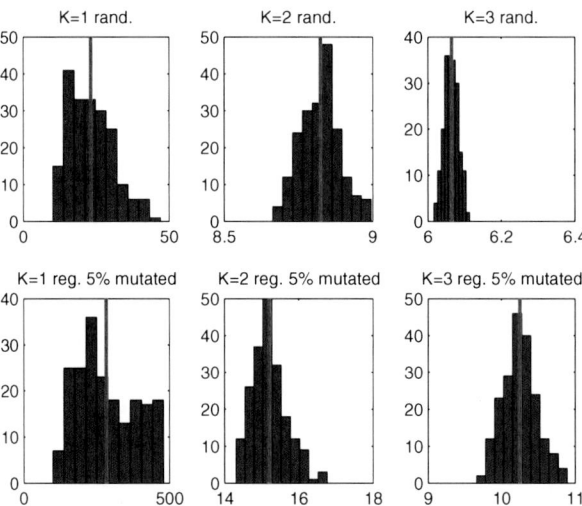

Fig. 2. Distributions of the average path lengths of networks from different ensembles.

of the parameters are distinct, for example in terms of their means, the variation within the ensembles is large enough that the parameters can not tie individual networks into ensembles unambiguously (Figures 7, 8).

For the dynamical parameters, $\lambda^{(5)}$ (a generalized measure of average sensitivity over five time steps) will capture the dynamical behavior of the networks. While this order parameter on average differs in different ensembles, it has a

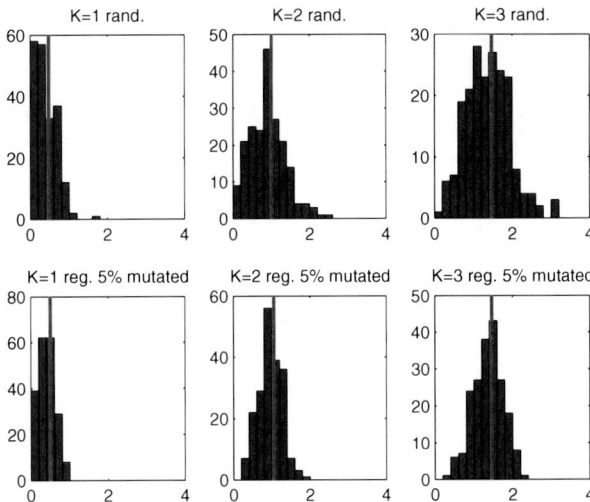

Fig. 3. Distributions of $\lambda^{(5)}$ for networks from different ensembles.

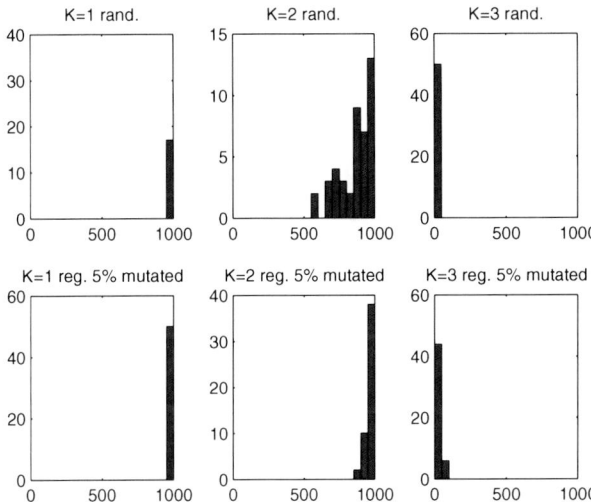

Fig. 4. Distributions of the size of frozen core for networks from different ensembles.

large variation in individual finite size network realizations from the ensembles. This is due to both variation between finite-size networks selected and variation in the sample paths selected for averaging. Size of the frozen component is more constant over different networks but lacks the separation power of different ensembles. In fact, most networks look similar in terms of this parameter.

Instead of computing individual statistics from networks, we should compare the entire networks directly. While there are several aspects that make this

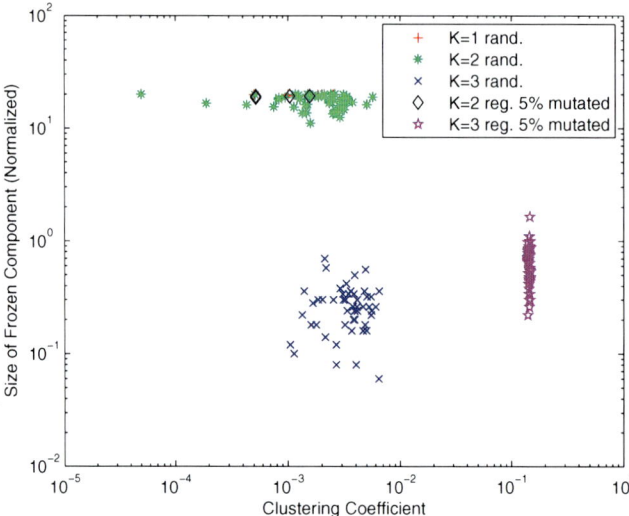

Fig. 5. Scatter plots of clustering coefficients and frozen core sizes for networks from different ensembles. For K=1 reg. 5% mutated networks clustering coefficient has a value zero and thus, this ensemble has been omitted from the illustration.

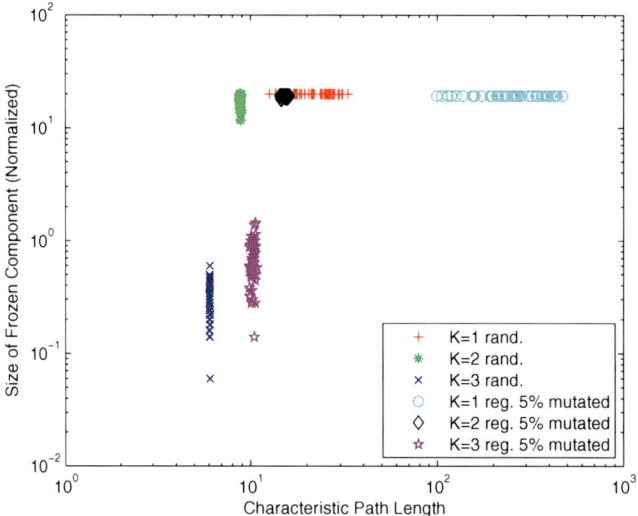

Fig. 6. Scatter plots of characteristic path lengths and frozen core sizes for networks from different ensembles.

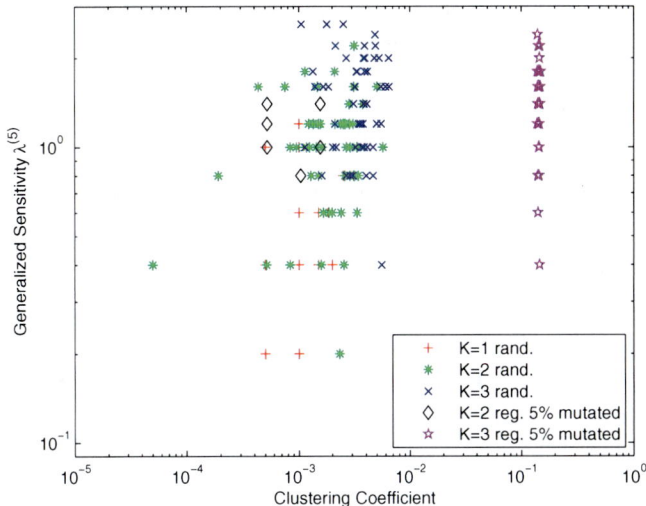

Fig. 7. Scatter plots of clustering coefficients and generalized sensitivity parameters $\lambda^{(5)}$ for networks from different ensembles. For K=1 reg. 5% mutated networks clustering coefficient has a value zero and thus, this ensemble has been omitted from the illustration.

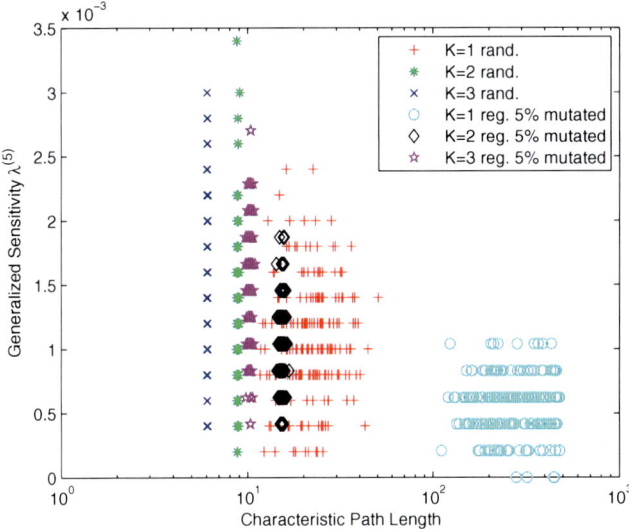

Fig. 8. Scatter plots of characteristic path lengths and generalized sensitivity parameters $\lambda^{(5)}$ for networks from different ensembles.

comparison difficult, for example a difference in the number of nodes and the degree distributions, we can do the comparison using the information theory based approach. Thus, we can compare the networks using their information content.

4.2 Information Theoretical Relationships

Making use of the universal information distance as approximated by the NCD, we are able to relate the structure of a network to its dynamics without reducing the network to an arbitrary set of features, thus allowing us to capture the information flow through the network in an unbiased (feature-independent) manner [43]. It also allows us to distinguish different ensembles of dynamical networks based solely on their structure, on their dynamics, or on a combination of both. To illustrate this, we generated six Boolean network ensembles ($N = 1000$) with two different wiring topologies: random and regular, each with $K = 1, 2$, or 3.

The Boolean network structure is represented by defining the wiring matrix W and the truth tables (functions) F. The wiring matrix contains N rows corresponding to the nodes in the network. Each row i contains the numbers of the nodes that are connected to node i. We extend this encoding further such that instead of using the absolute numbers for the nodes, we represent the connections by distances along an arbitrary linear arrangement of the nodes. For example, if node 20 is connected to node 8, then row 20 in W will contain a

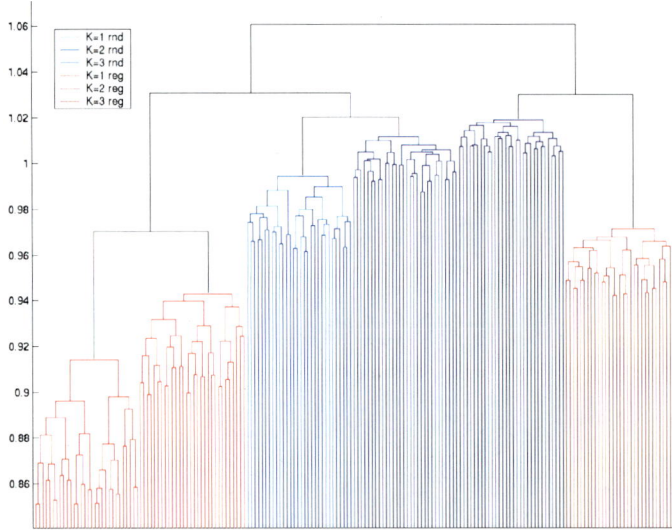

Fig. 9. The normalized compression distance (NCD) applied to all pairs of networks. The resulting distance matrix was then used to build a dendrogram, using the complete linkage method. Six ensembles of random Boolean networks ($K = 1, 2, 3$ each with random or regular topology; $N = 1000$) were used to generate 30 networks from each ensemble.

−12. This encoding is effective in the sense that the regularities in the network structure are easily observable. For example, if the network is regularly wired as in a cellular automaton, this is clearly visible in the connection matrix. Following the matrix representation of the connections, the truth tables F of the Boolean functions are given for each node i. The matrix F defines how the output of the node is obtained using the inputs defined in W.

In comparing the network structures, one distinctive parameter is their size and thus, the size of their respective compressed data structures. For example, using our encoding for the network structure, Boolean networks with connectivity $K = 1, 2$, and 3 could be separated from each other simply by looking at the sizes of W and F. To verify that the proposed method is able to find the information within the network structure and does not merely classify the networks based on their size, we introduce fictitious inputs to make all network representations to be of equal size. A fictitious input is one that does not affect the output of the function. We choose the fictitious input(s) randomly to be one of the existing connections; that is, the same connection is repeated more than once in the network encoding. For our analyzes, fictitious inputs have been applied such that all Boolean networks are encoded with the same size as $K = 3$ networks. As Figure 9 illustrates, all of the different ensembles considered are clearly distinguishable.

To extend this analysis further, we used NCD to study the relationship between structural information and dynamical behavior within a common framework. Within each of the above 6 Boolean network ensembles, we generated 150 networks and calculated the NCD between all pairs of network structures and between their associated dynamic state trajectories. In order to obtain comparable data from the dynamics of Boolean networks (that is, state-space trajectories) from different ensembles, we performed a burn-in of 100 time steps before collecting the data. This was done in order to ensure that the network is not in a transient state. Based on simulations, using a longer burn-in period did not affect the results. After the burn-in, trajectories were collected for 10 consecutive time steps, and subsequently were used to compute the NCD between pairs of trajectories. Using longer time series did not affect the results. This process was repeated for exactly the same networks that were used for comparing network structures.

The relationship between structure and dynamics was visualized by plotting the structure-based NCD versus the dynamics-based NCD for pairs of networks within each ensemble (Figure 10). All network ensembles were clearly distinguishable based on their structural and dynamical information. Additionally, the critical ensemble ($K = 2$, random wiring) exhibited a distribution that is markedly more elongated along the dynamics axis as compared to the chaotic and ordered ensembles, supporting the view that critical systems exhibit maximal diversity. The wide spread of points for the critical network ensemble in Figure 10 shows that their dynamics range between those of ordered and chaotic ensembles. Indeed, very different network structures can yield both relatively similar and dissimilar dynamics, thereby demonstrating the dynamic diversity

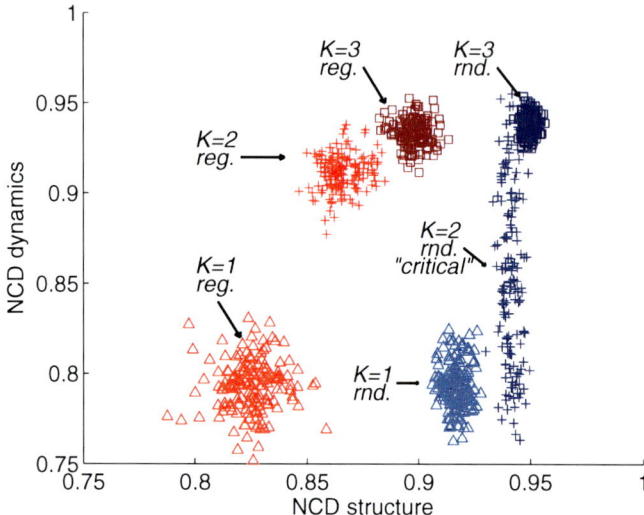

Fig. 10. The normalized compression distance (NCD) applied to network structure and dynamics. Six ensembles of random Boolean networks ($K = 1, 2, 3$ each with random or regular topology; $N = 1000$) were used to generate 150 networks from each ensemble. NCDs were computed between pairs of networks (both chosen from the same ensemble) based on their structure (x-axis) and their dynamic state trajectories (y-axis). Different ensembles are clearly distinguishable. The critical ensemble is more elongated, implying diverse dynamical behavior.

exhibited in the critical regime. Thus, the universal information distance provides clear evidence that the most complex relationships between structure and dynamics occur in the critical regime.

4.3 Quantifying Information Propagation

Dynamical behavior of a system can be characterized using an order parameter as discussed in section 3.2. This analysis can directly be generalized into more general information theoretic framework. Instead of using the Hamming distance as the measure of similarity, we are using the normalized information distance. In computational applications normalized compression distance can be used as an approximation. Thus, the information-based Derrida map is obtained by computing the distances between the states $s^{(1)}(t)$ and $s^{(2)}(t)$ using $d(t) = d_{NCD}(s^{(1)}(t), s^{(2)}(t))$ (Figure 11).

When compared with the traditional Derrida map for random Boolean networks, our information-based version has an interesting property. For a critical network the curve stays at the diagonal for all the distances, not just close to the origin. With the traditional approach that is based on the Hamming distance the dynamical regime can be characterized only by using very small perturbations, as the order parameter is defined by the slope at the origin. Our information-based

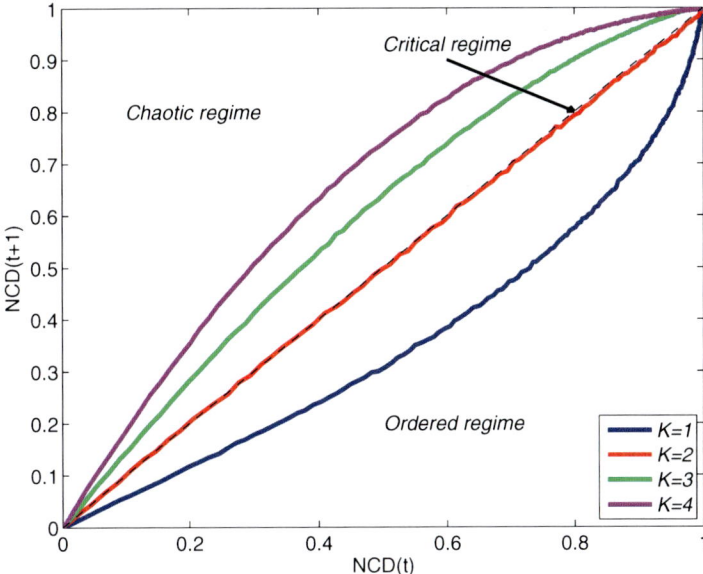

Fig. 11. Information-based Derrida map for Boolean networks from ensembles $K = 1, 2, 3, 4$ with $b = 0.5$. The dynamical regime can be observed throughout the curve.

version allows us to use perturbations of any size as the same dynamical behavior is observed throughout the curve. This allows us to apply the information-based order parameter directly to data, measured from real dynamical systems. For example, when a stimulus is given to a biological system, it is usually not known what the exact response is. Thus, our measure allows the usage of biological data even though the size of the response, or the perturbation, is not known. This type of analysis of information propagation is discussed in [44].

4.4 Set Complexity

Evident limitation in the presented NCD based analysis is that it inherently considers pairs of networks or pairs of trajectories. While this can reveal interesting insights, it makes interpretation of results challenging. A step towards more general analysis would be to quantify the information content of individual networks and their dynamics. For this purpose we would need to be able to quantify the effective information content of the network and its dynamics. Standard measures of information, such as entropy or Kolmogorov complexity, only quantify the randomness of the objects. For this type of application, we would need to apply context-dependent measure for information.

A recent development in information theory is a measure called set complexity (see Section 2.2). Here we show how this measure can be applied to quantify information in the network structure and dynamics.

Structure of a network can be represented in a form of adjacency graph $A^{n \times n}$, where n is the number of nodes in the network. We can then define a set S as the collection of the nodes of the network $S = \{A_1, A_2, \ldots, A_n\}$, where $A_i = [a_{i1}, a_{i2}, \ldots, a_{in}]$. $a_{ij} = 1$ if there is a connection from node i to node j. Thus, set S includes the rows (or columns) of the adjacency matrix as the strings. With this definition, we can directly apply the set complexity measure (Equation 9) to quantify the information embedded into network structure. An illustrative example on how set complexity behaves on different topological networks is shown in Figure 12.

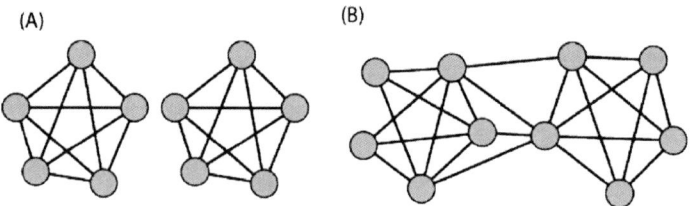

Fig. 12. Information content of two graphs with $N = 10$. Graph (A) has a low information content: $\Psi_A = 0.2$. Graph (B), the maximally informative undirected, unweighted graph with $N = 10$, on the other hand, has a much higher information content: $\Psi_B = 1.9$.

Dynamics can be directly be quantified by forming a set S as from the state vectors of the network over l consequtive time steps, $S = \{s(t), s(t+1), \ldots, s(t+l)\}$. To demonstrate this approach, we applied set complexity to state trajectories generated by ensembles of random Boolean networks operating in the ordered, chaotic, and critical regimes. Specifically, we have set the connectivity to be $k = 3$ and tuned the bias p in increments of 0.01 so that the average sensitivity, s, varies from $s < 1$ (ordered) to $s > 1$ (chaotic). For each value of s, 50 random networks (number of nodes, $n = 1000$) were each used to generate a trajectory of 20 states, after an initial "burn in" of running the network 100 time steps from a random initial state in order to allow the dynamics to stabilize (i.e., reach the attractors). We collected these 20 network states into a set for each network of the ensemble and calculated set complexity for each. Figure 13 shows the average set complexity over the 50 networks as a function of the average network sensitivity s (plotted a function of $\lambda = \log s$).

This example clear shows that set complexity can be used to extract relevant information from network dynamics and thus can be used to quantify dynamical behavior of the network. It is straightforward to link structure based and dynamics based set complexity trough correlation analysis. Thus, we can use set complexity to study structure dynamics relationships of individual networks without reducing them into topological or dynamical parameters.

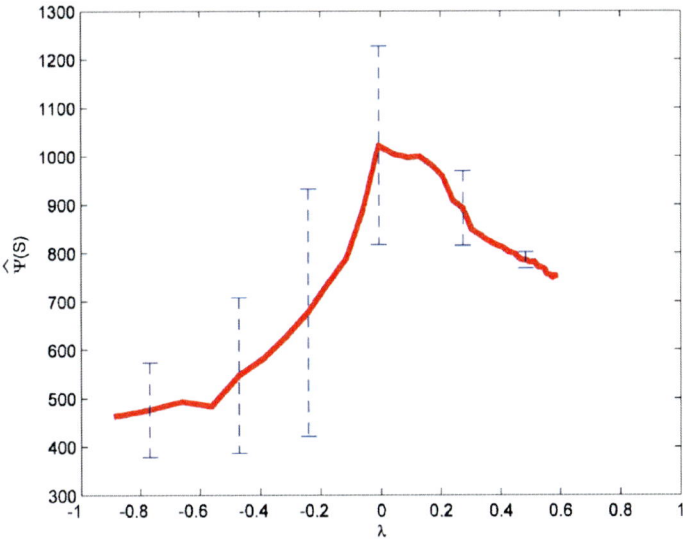

Fig. 13. The average, estimated set complexity of random state trajectories as a function of the log of the average sensitivity λ, the Lyapunov exponent, generated by networks operating in the ordered, critical, and chaotic regimes. The bars show the variability (one standard deviation) of the estimated set complexity for 50 networks [45].

5 Conclusions

In this chapter we have demonstrated how the structure dynamics relationships of the model networks can be studied. We reviewed the most common parameters for quantifying structure and dynamics of networks and used these to analyze various ensembles of Boolean networks. This analysis demonstrated some major limitations of this standard analysis.

To overcome these limitations and to make the analysis as general as possible, we proposed to use information theory, implemented in form of NCD, as a powerful framework for extracting structure-dynamics relationships. We demonstrated the potential of the proposed approach with multiple applications to the analysis of structure and dynamics. It is fascinating that the analyzes presented herein are possible using only the compressibility of a file encoding the network or its dynamics without needing to select any particular network parameters or features. This approach allows us to study, under a unified information theoretic framework, how a change in structural complexity affects the dynamical behavior, or vice versa.

Future work should work on developing more optimal analysis techniques that implements the information theoretical approach. For example, estimation of NCD can be further improved by developing context dependent compression algorithms. Furthermore, analysis should be extend to real data and to real biological networks. Here, various statistical tests are needed to quantify how well real data corresponds to model networks.

References

1. Barrow-Green, J.: Poincare and the Three Body Problem. AMS, Providence (1996)
2. Aldana, M., Cluzel, P.: A natural class of robust networks. Proceedings of the National Academy of Sciences USA 100(15), 8710–8714 (2003)
3. Klemm, K., Bornholdt, S.: Topology of biological networks and reliability of information processing. Proceedings of the National Academy of Sciences USA 102(51), 18414–18419 (2005)
4. Nochomovitz, Y.D., Li, H.: Highly designable phenotypes and mutational buffers emerge from a systematic mapping between network topology and dynamic output. Proceedings of the National Academy of Sciences USA 103(11), 4180–4185 (2006)
5. Variano, E.A., McCoy, J.H., Lipson, H.: Networks, dynamics, and modularity. Physical Review Letters 92(188701) (2004)
6. Shmulevich, I., Lähdesmäki, H., Dougherty, E.R., Astola, J., Zhang, W.: The role of certain Post classes in Boolean network models of genetic networks. Proceedings of the National Academy of Sciences USA 100(19), 10734–10739 (2003)
7. Albert, R., Jeong, H., Barabási, A.-L.: Error and attack tolerance of complex networks. Nature 406(6794), 378–382 (2000)
8. Barabási, A.-L., Albert, R.: Emergence of scaling in random networks. Science 286(5439), 509–512 (1999)
9. Watts, D.J., Strogatz, S.H.: Collective dynamics of 'small-world' networks. Nature 393(6684), 440–442 (1998)
10. Kauffman, S.A.: The ensemble approach to understand genetic regulatory networks. Physica A 340(4), 733–740 (2004)
11. Kauffman, S.A.: The Origins of Order: Self-organization and selection in evolution. Oxford University Press, New York (1993)
12. Cover, T.M., Thomas, J.A.: Elements of Information Theory, 2nd edn. Wiley-Interscience, Hoboken (2006)
13. Li, M., Vitanyi, P.: An Introduction to Kolmogorov Complexity and Its Applications, 2nd edn. Springer, New York (1997)
14. Kauffman, S.A.: Metabolic stability and epigenesis in randomly constructed genetic nets. Journal of Theoretical Biology 22, 437–467 (1969)
15. Derrida, B., Pomeau, Y.: Random networks of automata: a simple annealed approximation. Europhysics Letters 1, 45–49 (1986)
16. Shannon, C.E.: A mathematical theory of communication. Bell System Technical Journal 27, 379–423 (1948)
17. Kolmogorov, A.N.: Three approaches to the quantitative definition of information. Problems in Information Transmission 1(1), 1–7 (1965)
18. Solomonoff, R.: A formal theory of inductive inference. Information and Control 7, 1–22 (1964)
19. Chaitin, G.J.: On the length of programs for computing finite binary sequences: Statistical considerations. Journal of the Association of Computer Machinery 16(1), 145–159 (1969)
20. Cover, T.M., Thomas, J.A.: Elements of Information Theory. Wiley-Interscience, Hoboken (1991)

21. Krawitz, P., Shmulevich, I.: Basin entropy in Boolean network ensembles. Physical Review Letters 98(15), 158701(1-4)(2007)
22. Krawitz, P., Shmulevich, I.: Entropy of complex relevant components in Boolean networks. Physical Review E 76, 036115 (2007)
23. Huffman, D.A.: A method for the construction of minimum-redundancy codes. Proceedings of the Institute of Radio Engineers 40, 1098–1102 (1952)
24. Ziv, J., Lempel, A.: A universal algorithm for sequential data compression. IEEE Transactions on Information Theory 23(3), 337–343 (1977)
25. Rissanen, J., Langdon, G.G.: Arithmetic coding. IBM Journal of Research and Development 23, 149–162 (1979)
26. Bennett, C.H., Gacs, P., Li, M., Vitanyi, P.M.B., Zurek, W.: Information distance. IEEE Transactions on Information Theory 44(4), 1407–1423 (1998)
27. Li, M., Chen, X., Li, X., Ma, B., Vitanyi, P.: The similarity metric. IEEE Transactions on Information Theory 50(12), 3250–3264 (2004)
28. Cilibrasi, R., Vitanyi, P.: Clustering by compression. IEEE Transactions on Information Theory 51(4), 1523–1545 (2005)
29. Erdös, P., Rényi, A.: On random graphs. Publicationes Mathematicae 6, 290–297 (1959)
30. Albert, R., Barabási, A.-L.: Statistical mechanics of complex networks. Reviews of Modern Physics 74(1), 47–97 (2002)
31. Barabási, A.L.: Linked: The New Science of Networks. Perseus Books Group, Cambridge (2002)
32. Babu, M.M., Luscombe, N.M., Aravind, L., Gerstein, M., Teichmann, S.A.: Structure and evolution of transcriptional regulatory networks. Current Opinion in Structural Biology 14(3), 283–291 (2004)
33. Guelzim, N., Bottani, S., Bourgine, P., Képés, F.: Topological and causal structure of the yeast transcriptional regulatory network. Nature Genetics 31(60), 60–63 (2002)
34. Aldana, M., Coppersmith, S., Kadanoff, L.P.: Boolean dynamics with random couplings. In: Kaplan, E., Marsden, J.E., Sreenivasan, K.R. (eds.) Perspectives and Problems in Nonlinear Science. A Celebratory Volume in Honor of Lawrence Sirovich. Springer Applied Mathematical Sciences Series, pp. 23–89. Springer, New York (2003)
35. Derrida, B., Stauffer, D.: Phase transitions in two dimensional Kauffman cellular automata. Europhysics Letters 2, 739–745 (1986)
36. Kesseli, J., Rämö, P., Yli-Harja, O.: Iterated maps for annealed Boolean networks. Physical Review E 74, 046104 (2006)
37. Matache, M., Heidel, J.: A random Boolean network model exhibiting deterministic chaos. Physical Review E 69, 056214 (2004)
38. Andrecut, M., Ali, K.: Chaos in a simple Boolean network. International Journal of Modern Physics B 15(1), 17–23 (2001)
39. Andrecut, M.: Mean field dynamics of random Boolean networks. Journal of Statistical Mechanics P02003 (2005)
40. Rämö, P., Kesseli, J., Yli-Harja, O.: Stability of functions in gene regulatory networks. Chaos 15, 034101 (2005)
41. Shmulevich, I., Kauffman, S.: Activities and sensitivities in Boolean network models. Physical Review Letters 93(4), 048701(1-4) (2004)
42. Bernasconi, A.: Mathematical Techniques for the Analysis of Boolean Functions. PhD thesis, University of Pisa (1998)

43. Nykter, M., Price, N.D., Larjo, A., Aho, T., Kauffman, S.A., Yli-Harja, O., Shmulevich, I.: Critical networks exhibit maximal information diversity in structure-dynamics relationships. Physical Review Letters 100(058702) (February 2008)
44. Nykter, M., Price, N.D., Aldana, M., Ramsey, S., Kauffman, S.A., Hood, L., Yli-Harja, O., Shmulevich, I.: Gene expression dynamics in the macrophage exhibit criticality. Proceedings of the National Academy of Sciences USA 105(6), 1897–1900 (2008)
45. Galas, D., Nykter, M., Carter, G., Price, N.D., Shmulevich, I.: Set-based complexity and biological information. IEEE Transactions on Information theory 56(2), 667–677 (2010)

Large-Scale Statistical Inference of Gene Regulatory Networks: Local Network-Based Measures

Frank Emmert-Streib

Computational Biology and Machine Learning, Center for Cancer Research and Cell Biology, School of Medicine, Dentistry and Biomedical Sciences, Queen's University Belfast, 97 Lisburn Road, Belfast, BT9 7BL, UK
v@bio-complexity.com

Abstract. In this chapter we discuss various local network-based measures in order to assess the performance of inference algorithms for estimating regulatory networks. These statistical measures represent domain specific knowledge and are for this reason better adapted to problems that are directly involving networks compared to other measures frequently used in this context like the F-score. We are discussing three such measures with special focus on the inference of regulatory networks from expression data. However, due to the fact that currently there is a vast interest in network-based approaches in systems biology the presented measures may be also of interest for the analysis of a different type of large-scale genomics data.

1 Introduction

Understanding the dynamical behavior of molecular processes and their relation to the biological function of cells, tissues and ultimately of an organism is the goal of systems biology [1–4]. In contrast to molecular biology, which pursued a gene-centric view [5], systems biology aims at studying interacting components of molecular biological systems in order to understand the emerging phenotype [6, 7]. For this reason gene networks, e.g., the transcriptional regulatory, metabolic or protein networks [8–13] are of central importance for integrating information from various scales and variables [14–16]. Because gene networks represent biochemical interactions among gene products and not merely associations, they form causal [17, 18] instead of association networks [19]. This important role of gene networks explains the recent interest in estimating them from high-throughput data. Especially for expression data from DNA microarray experiments there have been many studies devoted to the estimation of regulatory networks [20–27]. It is generally assumed that these networks are not only useful to gain insights into normal cell physiology but also into pathological processes [6, 28, 29].

Despite the recent excitement about the perspectives opened by estimating regulatory networks on a genomic-scale there are considerable problems involved.

In this chapter we discuss one of these which relates to the assessment of the inference performance of estimation methods. So far, much attention has been devoted to the development of novel inference algorithms [30–35]. Interestingly, much less afford has been put into the invention of new analysis methods that consider the problem under investigation explicitly. The crucial point here is that usually general purpose measures from statistics are employed to analyze inference methods, e.g., precision, recall or the F-score. The reason these measures are generic is that they can be applied to any problem regardless of its nature. At first, this may seem as advantage, however, on the other side, these measures do not take domain knowledge into account. For example, in the context of inferring regulatory networks the data are apparently not exchangeable but are related in a structured manner. A F-score is a global measure that cannot be directly related to structural components of a network. For this reason it seems beneficial to devise statistical measures that are network-based.

In this chapter we review several local network-based measures recently introduced [36, 37]. We study these measures by applying four widely used inference algorithms, ARACNE [32], CLR [34], MRNET [33] and Relevance Networks (RN) [38], to expression data. All of these inference methods are based on information-theoretic measures [39, 40] which makes a comparison sensible.

This chapter is organized as follows. In the next section we describe the methods used and in section 3 we present numerical results. In section 4 we provide a discussion of local network-based measures and this chapter finishes in section 5 with conclusions.

2 Methods

The purpose of this chapter is to advocate and demonstrate the usage of local network-based measures in assessing inference algorithms [36, 37]. Specifically, we are focusing on the inference of regulatory networks. Commonly, network inference algorithms have been studied and their performance evaluated using measures like precision, recall, F-score or AUROC (area under the receiver operator characteristic) [33, 41, 42]. All these measures have in common that they assess the performance of an inference algorithm by one (scalar) value. With other words, the quality of an inferred regulatory network is globally evaluated by a value of one of the aforementioned measures. A problem with such a global evaluation is that these measures do not provide insights into the inference performance of structural regions of networks, e.g., motifs, subnetworks or modules, hubs or even individual edges. The reason for this shortcoming is that these measures do not make use of the domain of the problem but are multipurpose measures. In order to overcome this problem *local network-based* measures are suggested and their usage is demonstrated.

2.1 Definition of the Model

The *local network-based* measures we define in the following are based on ensemble data and the availability of a reference network that represents the 'true' regulatory network [36, 37]. Here we describe the approach by using simulated expression data and discuss extensions to microarray data in section 5.

The generation of ensemble data is visualized in Fig. 1. For a given structure of a regulatory network G, E simulated expression data sets, $\mathcal{D} = \{D_1(G), \ldots, D_E(G)\}$, are generated by using E different parameter sets. Each set of parameters defines the dynamics of expression values mimicking the dynamics of gene regulation [43, 44]. More precisely, for the following analysis we use a subnetwork of the transcriptional regulatory network of yeast [9], G, consisting of 100 genes. This subnetwork was randomly sampled from the entire transcriptional regulatory network by using SynTReN [45]. From this network, expression data in steady-state condition are obtained by using dynamic equations with Michaelis-Menten and Hill enzyme kinetics. These data are generated with SynTReN [45]. In total, we generate $E = 300$ different data sets for sample size 200 and another $E = 300$ data sets for sample size 20 using the same subnetwork G of the transcriptional regulatory network of yeast but with different kinetic parameters for each data set. Biologically, the data $\mathcal{D}$ may correspond to a population of one species spanning the whole dynamic range different individual organisms from the same species can exhibit. The reason for this variability comes from the fact that molecular systems behave unlike a clockwork utilizing parallel pathways for inter- and intra-cell communication. Statistically, using one network structure G underlying the ensemble data allows the statistical assessment of network components down to the level of individual edges. If different networks, $G_i \neq G_j$ for $i \neq j$, would be used for different data sets $D_i(G_i)$ for $i \in \{1, \ldots, E\}$, the identification of such network components would be no longer possible and, hence, an averaging over different data sets would become meaningless. For example, given two networks of the same size, there may be an edge connecting gene m and n in the first network but no edge in the second network. This demonstrates the problem to identify common parts in these networks. If instead of mathematical labels, m and n, the nodes in these networks would be labeled with gene names, this problem would become more apparent.

After having obtained an ensemble of estimated networks $\mathcal{G}^e = \{G_i^e\}_{i=1}^E$ from $\mathcal{D} = \{D_1(G), \ldots, D_E(G)\}$ and the application of an inference algorithm we can summarize the result in various ways. The simplest possible way is illustrated in Fig. 2. For a given regulatory network G, we assume to be true, we can derive two representations. One we call the *TP network* and another the *TN network*. The TP network contains edges between two genes if these genes are directly connected in the regulatory network, whereas in the TN network two genes are connected if there is no direct link between these genes. Hence, the TP network represents *true positive* edges, whereas the TN network represents *true negative* connections. From $\mathcal{G}^e = \{G_i^e\}_{i=1}^E$ one can estimate the TPR (true positive rate), respectively, the TNR (true negative rate) of these edges, which can be seen as the weights of edges, hence, leading to undirected but weighted networks. This is illustrated in Fig. 2 with a simple numerical example. Based on $\{TPR_{ij}\}$ and $\{TNR_{ij}\}$ one can derive various measures in order to assess the performance of an inference algorithm. In the following we present three such measures that

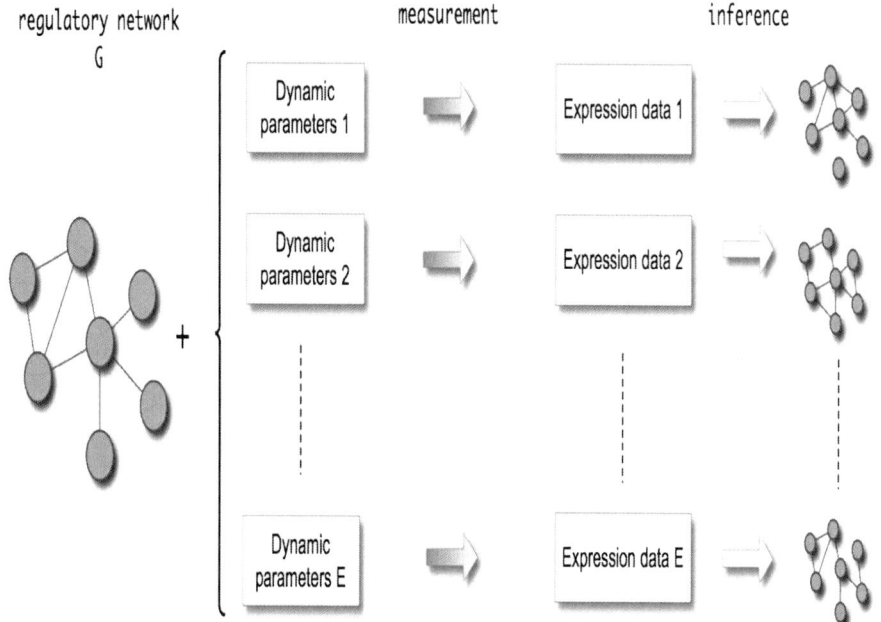

Fig. 1. Illustration of our simulation set up. One transcriptional regulatory network G is used to generate E different data sets. These data sets are obtained by a variation of the dynamical parameters of the dynamical system used to simulate gene expression. To each of the E data sets an inference algorithm is appled to infer a network.

reflect a local structural component of the network. In this respect these entities are local network-based measures. We would like to emphasize that there are many more measures possible.

Principally, any combination of $\{TPR_{ij}\}$ and $\{TNR_{ij}\}$ values would result in a valid (statistical) measure, however, we are focusing on measures that may be especially beneficial for the analysis of regulatory networks. For this reason, we present in the following three measures that allow a sensible biological interpretation. Due to the fact that biologists are traditionally interested in the behavior of individual genes [5], TP and TN rates of individual edges are useful in this context. This represents also the simplest possible measure consisting of individual values of $\{TPR_{ij}\}$ and $\{TNR_{ij}\}$ only.

The second measure is based on three-gene motifs. Recently, it has been recognized that motifs are important building blocks of various complex networks including biological ones [46–51]. For this reason, biologically, it may be of interest to study their inferability. We define the true reconstruction rate of a motif by

$$p = \frac{1}{3} \sum_{1}^{3} TXR_i. \tag{1}$$

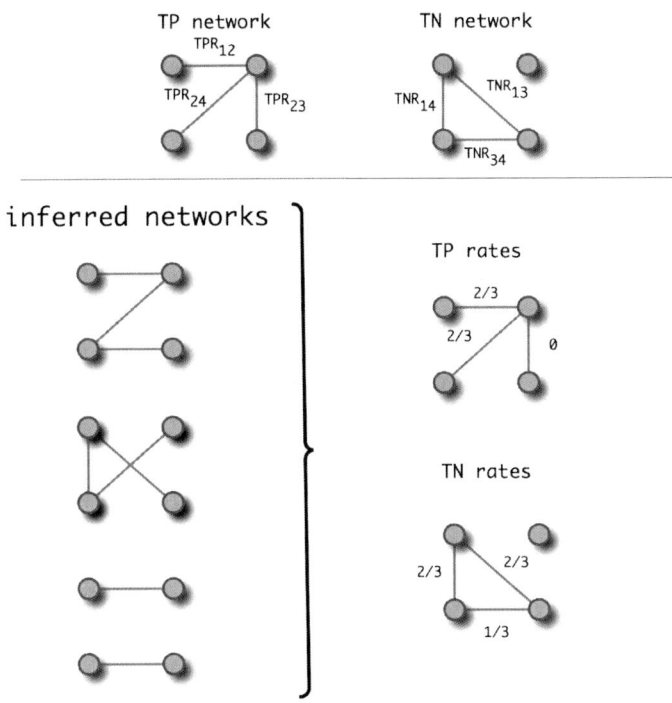

Fig. 2. Illustration of the summary of $\mathcal{G}^e = \{G_i^e\}_{i=1}^E$. A true network G can be split into two weighted networks, called TP and TN network. The weights of these networks are obtained from $\mathcal{G}^e$. A simple example for $E = 3$ illustrates the procedure.

Here TXR corresponds either to a TPR if two genes are connected or to TNR if these genes are unconnected and the factor results form the fact that we consider three-gene motifs only. Fig 3 illustrates all five three-gene motifs that are possible. These motifs are directed because the transcriptional regulatory network is a directed network. From Eqn. 1 follows for the five motifs

$$p[\text{motif} = 1] = \frac{1}{3}\big(\text{TPR}(a \to b) + \text{TPR}(b \to c) + \text{TNR}(a \nleftrightarrow c)\big), \qquad (2)$$

$$p[\text{motif} = 2] = \frac{1}{3}\big(\text{TPR}(a \to b) + \text{TPR}(b \leftarrow c) + \text{TNR}(a \nleftrightarrow c)\big), \qquad (3)$$

$$p[\text{motif} = 3] = \frac{1}{3}\big(\text{TPR}(a \leftarrow b) + \text{TPR}(b \to c) + \text{TNR}(a \nleftrightarrow c)\big), \qquad (4)$$

$$p[\text{motif} = 4] = \frac{1}{3}\big(\text{TPR}(a \to b) + \text{TPR}(b \to c) + \text{TPR}(a \to c)\big), \qquad (5)$$

$$p[\text{motif} = 5] = \frac{1}{3}\big(\text{TPR}(a \to b) + \text{TPR}(b \to c) + \text{TPR}(a \leftarrow c)\big). \qquad (6)$$

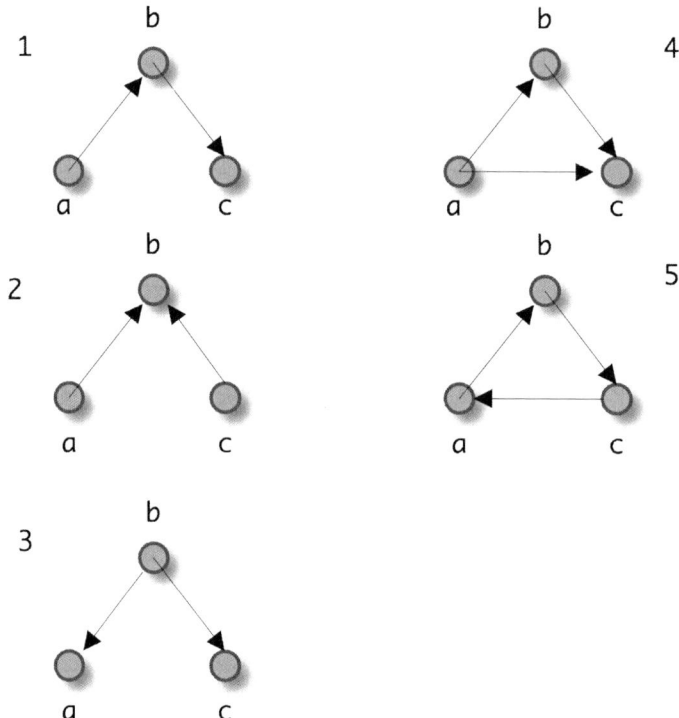

Fig. 3. Visualization of three-gene motifs. The motifs are directed because the underlying transcriptional regulatory network is a directed network. For this reason, there are only five different types of motifs consisting of three genes.

Each of these five measures represents the inferability of a certain motif type. Principally, one could characterize every motif found in the transcriptional regulatory network in this way. However, in the following we average over motifs of the same type found in the transcriptional regulatory network leading to the mean true reconstruction rate $\overline{p[\text{motif}]}$ of a certain motif type. The reason why we prefer to average over all motifs of the same type is that this leads to a considerable simplification giving, e.g., just one value of $\overline{p[\text{motif} = 3]}$ for a regulatory network instead of thousands (see results section).

Finally, the third measure we consider is related to activator and repressor edges in the transcriptional regulatory network. Here a link between two genes is called an activator edge if the regulatory effect between these two genes is positive and repressor edge if this effect is negative. This leads to a classification of all edges into two classes. Figure 4 illustrates this classification. Edges belonging to a class are either represented as bold, full line or thin, dashed line. From this classification according to the regulatory effect of a link between two genes we obtain two sets of TPRs, $\mathcal{A} = \{TRP_{ij} | e_{ij} \text{ is activator edge}\}$ and $\mathcal{R} = \{TRP_{ij} | e_{ij} \text{ is repressor edge}\}$. From these two sets one can derive various

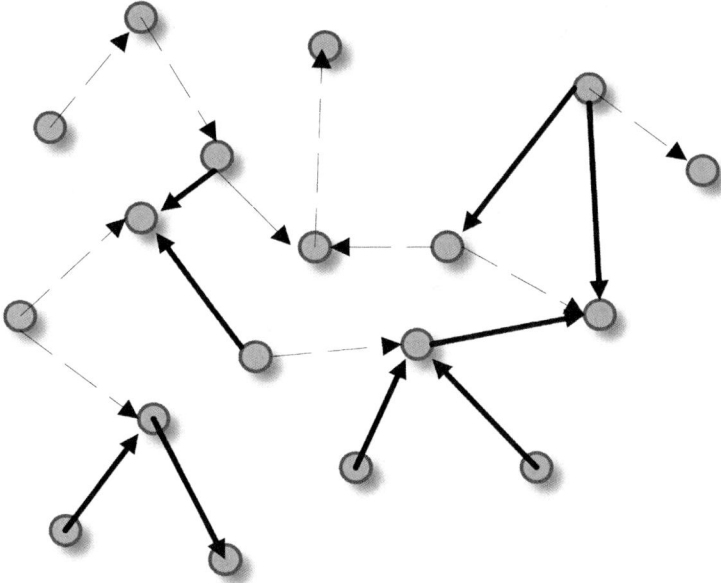

Fig. 4. A simplified network consisting of two types of edges. One type is represented by full, bold lines and the other type by dashed, thin lines.

measures, e.g., their mean or median value. However, in our analysis we want to compare these sets with respect to the distribution of TPRs of activator and repressor edges.

We demonstrate the usage of these measures by studying four different inference algorithms, ARACNE, CLR, MRNET and Relevance Networks (RN). For detailed information about these GRNI algorithms we refer the readers to [32–34, 38, 52]. For ARACNE we set the DPI (data processing inequality [39]) tolerance parameter $\epsilon = 0.1$, as in [53]. The MI values are estimated using non-parametric Gaussian estimator as described in [52] and [54]. The optimal cut-off value for each data set, D_i, used to declare edges significant is obtained by maximizing the F-score,

$$F(I_0') = \frac{2p(I_0')r(I_0')}{p(I_0') + r(I_0')}. \tag{7}$$

Here the F-score, $F(I_0')$, precision,

$$p(I_0') = TP/(TP + FP), \tag{8}$$

and recall,

$$r(I_0') = TP/(TP + FN), \tag{9}$$

are a function of the MI threshold I_0' (CLR uses z scores instead [34]) and so are the number of *true positive* (TP), false positive (FP) and false negative (FN) edges. This results in E different F-scores correspondingly E inferred networks for the ensemble $\mathcal{D} = \{D_1(G), \ldots, D_E(G)\}$ of a given sample size.

3 Results

In the following we demonstrate the usage and utility of the local network-based performance measures described in the previous section [36, 37].

3.1 Activator vs. Repressor Connections

The first measure we apply allows to investigates the influence of activator (positive effect) and repressor (negative effect) edges on their inferability. In Fig. 5 we show histograms to visualize the effect of activator (red) and repressor (blue) edges on the true positive rate (TPR) of edges for four GRNI algorithms (Top: ARACNE (left) and CLR (right). Bottom: MRNET (left) and RN (right).). The sample size for these data was 200.

The TPR of an edge is the number of times a specific edge is inferred correctly divided by the total number of data sets (E). To investigate the results in Fig. 5 quantitatively we apply a two-sample Kolmogorov-Smirnov test [55, 56], to each GRNI algorithm, testing for differences in the cumulative distribution function (CDF) of activator and repressor edges. For sample size 200 we obtain p-values of

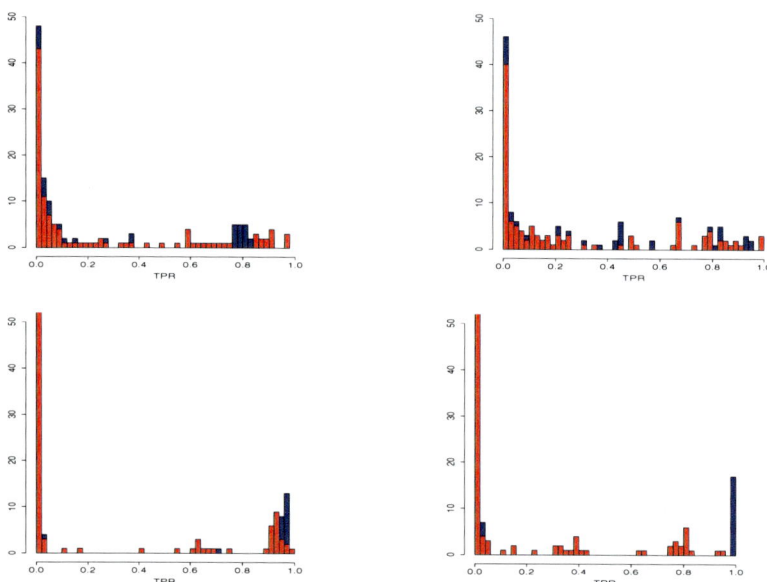

Fig. 5. Histogram of true positive rates for edges in the true network. Top: ARACNE (left) and CLR (right). Bottom: MRNET (left) and RN (right). Red indicates the contribution from activator and blue from repressor edges. Sample size is for all figures 200. (Reproduced by permission of Oxford University Press.)

0.0009688, 0.001554, 0.0001145, 3.432×10^{-6} for ARACNE, CLR, MRNET and RN respectively. The results suggest that, for a significance level of $\alpha = 0.01$, the edge type has a systematic effect on all four inference algorithms. Qualitatively, this can be seen from the histograms in Figure 5 where the repressor edges have a higher TPR, which means that they are easier to infer. We repeated the same analysis for a sample size of 20 and obtained p-values of 0.1097, 0.00823, 0.08638, 4.726×10^{-5} for ARACNE, CLR, MRNET and RN respectively. These results indicate that only CLR and RN are systematically affected by the edge type. In summary, this means that not only the used inference algorithm may introduce a bias in this context but also the sample size.

3.2 Motif Types

The next local network-based measure evaluates the inferability of basic motif types consisting of three genes, as shown in Fig. 3. In Table 1 we present the results for four of these network motifs for sample size 200, providing their mean true reconstruction rate $\bar{p}$ and its standard deviation $\sigma(\bar{p})$, $\#m$ corresponds to the number of motifs found in the network. The reason why we present only results for four of the five possible network motifs is that the network structure used in our analysis did not include cycles (see motif of type 5 in Fig. 3). Here the mean true reconstruction rate is calculated according to Eqn. 2-5.

From the table we can observe that all algorithms behave similarly with respect to the inferability for these network motifs, favoring motif type 1 and 3. The mean true reconstruction rate for motif type 4 is consistently the worst for all four methods. A general explanation for this behavior (different $\bar{p}$ values for different motif types) seems not to be straightforward. However, assuming

Table 1. Summary of motif statistics for ARACNE, CLR, MRNET and RN. (Reproduced by permission of Oxford University Press.)

	measure/motif type	1	2	3	4
ARA	$\#m$	40	171	446	10
	$\bar{p}$	0.591	0.352	0.530	0.156
	$\sigma(\bar{p})$	0.15	0.04	0.18	0.12
CLR	$\#m$	40	171	446	10
	$\bar{p}$	0.506	0.378	0.480	0.171
	$\sigma(\bar{p})$	0.131	0.072	0.137	0.194
MR	$\#m$	40	171	446	10
	$\bar{p}$	0.568	0.326	0.582	0.176
	$\sigma(\bar{p})$	0.1576	0.0083	0.2237	0.1434
RN	$\#m$	40	171	446	10
	$\bar{p}$	0.511	0.321	0.515	0.151
	$\sigma(\bar{p})$	0.121	0.010	0.152	0.112
CLR (EC)	$\#m$	2105	321896	1315	997
	$\bar{p}$	0.3625	0.3353	0.3355	0.0558
	$\sigma(\bar{p})$	0.103	0.026	0.031	0.168

a homogeneous TPR for an edge of 0.17 (as implied by motif type 4) and an idealized TNR of 1.0 for a non-edge would result in, e.g., for motif type 1 or 2, $\bar{p} = 0.45 = (2 \times 1.0 + 0.17)/3$. This value would be an upper bound that can not be exceeded because a TNR of 1.0 is the highest possible value. However, as one can see from table 1 all $\bar{p}$ values for motif type 1 and 3 are larger than 0.45. Hence, the actual situation is more intricate implying a complex dependency structure. For sample size 20 (not shown) we observe results that are qualitatively similar to the results discussed above. These results suggest that there is no bias introduced by the inference algorithms regarding the inferability of individual motif types (compare columns in table 1), however, each algorithm is biased towards motif types 1 and 2 (compare rows in table 1).

3.3 Individual Edges

The third measure we investigate studies the inferability of individual edges. This corresponds to the finest resolution a local network-based measure can assume but it gives also the most complex information.

In order to organize the complexity arising from the amount of information provided by the edge-specific inferability we present these results graphically. In Fig. 6 we show a visualization of the mean TPR of edges mapped onto the true regulatory network. These results have been obtained for CLR - corresponding results for ARACNE, MRNET and RN can be in found in [37]. The color code of the edges corresponds to their mean TPR. Specifically, for black edges, $1 \geq \overline{TPR} > 0.75$, for blue edges, $0.75 \geq \overline{TPR} > 0.5$, for green edges, $0.5 \geq \overline{TPR} > 0.25$, and for red edges, $0.25 \geq \overline{TPR} \geq 0.0$. A visual inspection of these figures suggests that there might be a systematic influence of *in-hubs* and *leafs* on the inferability as reflected by the color of edges. Here an in-hub is defined as a gene that has more then 3 incoming edges. We term these incoming edges as in-hub edges. A leaf node is a terminal gene that has exactly one incoming edge. We call this edge a leaf edge. A quantitative analysis of these observations for ARACNE, CLR, MRNET and RN can be found in [37]. As a result we find that in general the probability to observe blue or black leaf edges is much higher than to observe red or green leaf edges whereas for the in-hub edges this situation is reversed. This implies a systematic bias for all four inference algorithms.

4 Discussion

The conceptional idea that motivated the introduction of local network-based measures is to categorize network *components* according to a given rule. For example, we could categorize edges in different classes according to a graph-theoretical description that is solely based on the network topology of the regulatory network. Or we could categorize edges according to their effect as defined by the dynamical equations generating the expression data. Three examples of such measures are discussed in this chapter but it should be clear that there are many more measures that can be constructed based on our motivating idea.

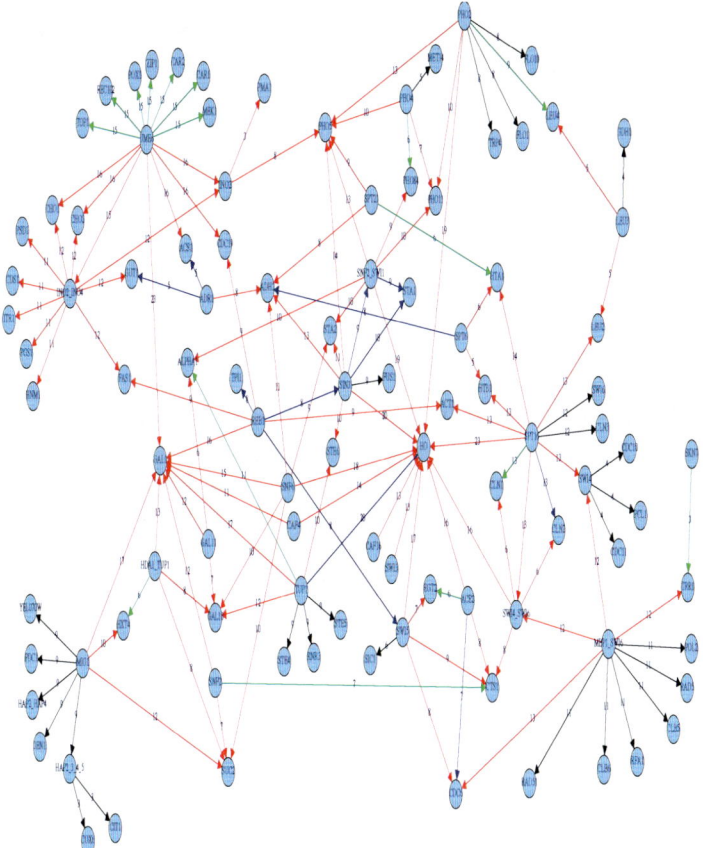

Fig. 6. Visualization of the results for CLR for sample size 200 inferring a subnetwork of yeast consisting of 100 genes. The color of each edge reflects its mean TPR. Specifically, for black edges, $1 \geq \overline{TPR} > 0.75$, for blue edges, $0.75 \geq \overline{TPR} > 0.5$, for green edges, $0.5 \geq \overline{TPR} > 0.25$, and for red edges, $0.25 \geq \overline{TPR} \geq 0.0$. The integer numbers at the edges correspond to the value of D_s, see [37]. (Reproduced by permission of Oxford University Press.)

More formal details about such measures and their defining rules can be found in [36, 37].

As the results from the previous section demonstrate, local network-based measures provide insights into the working mechanism of an inference algorithm beyond measures like precision or the F-score. This is no surprise because local network-based measures are *domain specific* rather than applicable to any statistical problem. Their utility depends on the context and, hence, the biological problem under investigation.

In addition to these results we think that local network-based measures can serve another purpose. Due to the fact that our understanding of molecular processes is very limited it is frequently difficult even to raise a hypotheses about the functioning of such processes. For this reason, means to generate hypotheses are of the utmost importance in order to create testable statements. In this respect local network-based measures may serve as exploratory analysis tools [57, 58]. But, for these measures to be useful, one needs to study the potential bias inference algorithms may cause carefully because otherwise an effect may be related to the inference algorithm rather than the underlying biological process. However, if this is addressed properly for a specific local measure of interest novel hypothesis may be found. Examples for this approach can be found in [36, 37] with respect to expression data from B-cells [53] but also for the design of statistical inference algorithms themselves.

In this chapter we analyzed local network-based measures by using simulated expression data. The reason for using simulated expression data instead of data from microarray experiments is that this way we could generate large sample sizes that are rarely available from microarray experiments. This way we could study the working mechanisms of different inference algorithms instead of revealing finite size effects that relate back to an insufficient amount of data. Principally, our measures are not limited to simulated data but can also applied to biological data. However, from a technical point of view we need an ensemble of data sets, instead of a single one. It is clear that this is a practical problem because usually no replicated data sets are generated if the sample size is in the hundreds, as is necessary for the inference of regulatory networks. A potential solution to this problem could be found by generating an ensemble of bootstrap samples from one data set [59–61]. This way one could circumvent the demanding requirements of our local network-based measures. In the future we will investigate this extension numerically.

5 Conclusions

With the advent of high-throughput data biology has been transformed into a technological field within a few years. As a side effect of this development there is a great demand for statistical and computational analysis methods that can cope with different types of genomics data and their integration [28, 62–66]. Despite the fact that we discussed local network-based measures for expression data, we are of the opinion that they may also be of use for other types of data, provided these data aim for the reconstruction of gene networks of any kind.

Acknowledgment

I would like to thank Gokmen Altay for fruitful discussions and Andrea Califano and members of his group for helpful with ARACNE. Some figures and tables of this chapter have been reproduced from [37] by permission of Oxford University Press (License Number 250436138947).

Funding: This project is supported by the Department for Employment and Learning through its "Strengthening the all-Island Research Base" initiative.

References

von Bertalanffy, L.: The theory of open systems in physics and biology. Science, 23–29 (1950)

Kauffman, S.: Metabolic stability and epigenesis in randomly constructed genetic nets. Journal of Theoretical Biology 22, 437–467 (1969)

Kauffman, S.: Origins of Order: Self-Organization and Selection in Evolution. Oxford University Press, Oxford (1993)

Waddington, C.: The strategy of the genes. Geo, Allen & Unwin, London (1957)

Beadle, G.W., Tatum, E.L.: Genetic Control of Biochemical Reactions in Neurospora. Proceedings of the National Academy of Sciences of the United States of America 27(11), 499–506 (1941)

Schadt, E.: Molecular networks as sensors and drivers of common human diseases. Nature 461, 218–223 (2009)

Vidal, M.: A unifying view of 21st century systems biology. FEBS Letters 583(24), 3891–3894 (2009)

Barabasi, A.L., Oltvai, Z.N.: Network biology: Understanding the cell's functional organization. Nature Reviews 5, 101–113 (2004)

Guelzim, N., Bottani, S., Bourgine, P., Kepes, F.: Topological and causal structure of the yeast transcriptional regulatory network. Nature Genetics (2002)

Lee, T.I., et al.: Transcriptional regulatory networks in *saccharomyces cerevisiae*. Science 298(5594), 799–804 (2002)

Maslov, S., Sneppen, K.: Specificity and Stability in Topology of Protein Networks. Science 296(5569), 910–913 (2002)

Newman, M.E.J.: The structure and function of complex networks. SIAM Review 45, 167–256 (2003)

Ravasz, E., Somera, A.L., Mongru, D.A., Oltvai, Z.N., Barabasi, A.L.: Hierarchical organization of modularity in metabolic networks. Science 297, 1551–1555 (2002)

Hwang, D., Rust, A., Ramsey, S., Smith, J., Leslie, D., Weston, A., de Atauri, P., Aitchison, J., Hood, L., Siegel, A., Bolouri, H.: A data integration methodology for systems biology. Proc. Natl. Acad. Sci. USA 102(48), 17296–17301 (2005)

Kitano, H.: Foundations of Systems Biology. MIT Press, Cambridge (2001)

Palsson, B.: Systems Biology. Cambridge University Press, Cambridge (2006)

Pearl, J.: Causality: Models, Reasoning, and Inference, Cambridge (2000)

Shipley, B.: Cause and Correlation in Biology. Cambridge University Press, Cambridge (2000)

Zhang, B., Horvath, S.: A general framework for weighted gene co-expression network analysis. Stat. Appl. Genet. Mol. Biol. 4, 17 (2005)

Emmert-Streib, F., Dehmer, M. (eds.): Analysis of Microarray Data: A Network Based Approach. Wiley-VCH, Chichester (2008)

de la Fuente, A., Bing, N., Hoeschele, I., Mendes, P.: Discovery of meaningful associations in genomic data using partial correlation coefficients. Bioinformatics 20(18), 3565–3574 (2004)

Lee, W.P., Tzou, W.S.: Computational methods for discovering gene networks from expression data. Brief Bioinform 10(4), 408–423 (2009)

Margolin, A., Califano, A.: Theory and limitations of genetic network inference from microarray data. Ann. NY Acad. Sci. 1115, 51–72 (2007)

Perrin, B.E., Ralaivola, L., Mazurie, A., Bottani, S., Mallet, J., d'Alche Buc, F.: Gene networks inference using dynamic Bayesian networks. Bioinformatics 19(2), ii138–ii148 (2003)

Stolovitzky, G., Monroe, D., Califano, A.: Dialogue on reverse-engineering assessment and methods: the dream of high-throughput pathway inference. Ann. NY Acad. Sci. 1115, 1–22 (2007)

Werhli, A., Grzegorczyk, M., Husmeier, D.: Comparative evaluation of reverse engineering gene regulatory networks with relevance networks, graphical gaussian models and bayesian networks. Bioinformatics 22(20), 2523–2531 (2006)

Wille, A., Bühlmann, P.: Low-order conditional independence graphs for inferring genetic networks. Statistical Applications in Genetics and Molecular Biology 4(1), 32 (2006)

Emmert-Streib, F., Dehmer, M. (eds.): Medical Biostatistics for Complex Diseases. Wiley-Blackwell, Chichester (2010)

Emmert-Streib, F.: The chronic fatigue syndrome: A comparative pathway analysis. Journal of Computational Biology 14(7), 961–972 (2007)

Butte, A., Tamayo, P., Slonim, D., Golub, T., Kohane, I.: Discovering functional relationships between rna expression and chemotherapeutic susceptibility using relevance networks. Proc. Natl. Acad. Sci. USA 97(22), 12182–12186 (2000)

Friedman, N.: Inferring cellular networks using probabilistic graphical models. Science 303(5659), 799–805 (2004)

Margolin, A., Nemenman, I., Basso, K., Wiggins, C., Stolovitzky, G., Dalla Favera, R., Califano, A.: Aracne: an algorithm for the reconstruction of gene regulatory networks in a mammalian cellular context. BMC Bioinformatics 7, S7 (2006)

Meyer, P., Kontos, K., Bontempi, G.: Information-theoretic inference of large transcriptional regulatory networks. EUROSIP journal on bioinformatics and systems biology 2007, 079879 (2007)

Faith, J.J., Hayete, B., Thaden, J.T., Mogno, I., Wierzbowski, J., Cottarel, G., Kasif, S., Collins, J.J., Gardner, T.S.: Large-scale mapping and validation of escherichia coli transcriptional regulation from a compendium of expression profiles. PLoS Biol. 5 (January 2007)

Xing, B., van der Laan, M.: A causal inference approach for constructing transcriptional regulatory networks. Bioinformatics 21(21), 4007–4013 (2005)

Emmert-Streib, F., Altay, G.: Local network-based measures to assess the inferability of different regulatory networks. IET Systems Biology 4(4), 277–288 (2010)

Altay, G., Emmert-Streib, F.: Revealing differences in gene network inference algorithms on the network-level by ensemble methods. Bioinformatics 26(14), 1738–1744 (2010)

Butte, A., Kohane, I.: Mutual information relevance networks: Functional genomic clustering using pairwise entropy measurements. In: Pacific Symposioum on Biocomputing, vol. 5, pp. 415–426 (2000)

Cover, T., Thomas, J.: Information Theory. John Wiley & Sons, Inc., Chichester (1991)

Gallager, R.: Information Theory and Reliable Communication. Wiley, Chichester (1968)

Fawcett, T.: An introduction to roc analysis. Pattern Recognition Letters 27, 861–874 (2006)

Husmeier, D.: Sensitivity and specificity of inferring genetic regulatory interactions from microarray experiments with dynamic bayesian networks. Bioinformatics 19(17), 2271–2282 (2003)

Albers, C., Jansen, R., Kok, J., Kuipers, O., van Hijum, S.: Simage: simulation of dna-microarray gene expression data. BMC Bioinformatics 7, 205 (2006)

Ribeiro, A., Zhu, R., Kauffman, S.: A general modeling strategy for gene regulatory networks with stochastic dynamics. Journal of Computational Biology 13(9), 1630–1639 (2006)

Van den Bulcke, T., Van Leemput, K., Naudts, B., van Remortel, P., Ma, H., Verschoren, A., De Moor, B., Marchal, K.: Syntren: a generator of synthetic gene expression data for design and analysis of structure learning algorithms. BMC Bioinformatics 7(1), 43 (2006)

Alon, U.: An Introduction to Systems Biology: Design Principles of Biological Circuits. Chapman & Hall/CRC, Boca Raton (2006)

Artzy-Randrup, Y., Fleishman, S.J., Ben-Tal, N., Stone, L.: Comment on "Network Motifs: Simple Building Blocks of Complex Networks" and "Superfamilies of Evolved and Designed Networks". Science 305(5687), 1107c (2004)

Dehmer, M., Emmert-Streib, F. (eds.): Analysis of Complex Networks: From Biology to Linguistics. Wiley-VCH, Chichester (2009)

Ciriello, G., Guerra, C.: A review on models and algorithms for motif discovery in protein-protein interaction networks. Brief Funct. Genomic Proteomic, eln015 (2008)

Milo, R., Shen-Orr, S., Itzkovitz, S., Kashtan, N., Chklovskii, D., Alon, U.: Network motifs: simple building blocks of complex networks. Science 298(5594), 824–827 (2002)

Shen-Orr, S., Milo, R., Mangan, S., Alon, U.: Network motifs in the transcriptional regulatory network of *Escherichia coli*. Nat. Genet. 31, 64–68 (2002)

Meyer, P., Lafitte, F., Bontempi, G.: minet: A r/bioconductor package for inferring large transcriptional networks using mutual information. BMC Bioinformatics 9(1), 461 (2008)

Basso, K., Margolin, A.A., Stolovitzky, G., Klein, U., Dalla-Favera, R., Califano, A.: Reverse engineering of regulatory networks in human b cells. Nature Genetics 37(4), 382–390 (2005)

Olsen, C., Meyer, P., Bontempi, G.: On the impact of entropy estimator in transcriptional regulatory network inference. EURASIP Journal on Bioinformatics and Systems Biology 2009, 308959 (2009)

Conover, W.: Practical Nonparametric Statistics. John Wiley & Sons, New York (1999)

Sheskin, D.J.: Handbook of Parametric and Nonparametric Statistical Procedures, 3rd edn. RC Press, Boca Raton (2004)

Hoaglin, D., Mosteller, F., Tukey, J.: Understanding Robust and Exploratory Data Analysis. Wiley, New York (1983)

Tuckey, J.: Exploratory Data Analysis. Addison-Wesley, Reading (1977)

Davison, A., Hinkley, D.: Bootstrap Methods and Their Application. Cambridge University Press, Cambridge (1997)

Efron, B.: Nonparametric estimates of standard error: The jackknife, the bootstrap and other methods. Biometrika 68(3), 589–599 (1981)

Efron, B., Tibshirani, R.: An Introduction to the Bootstrap. Chapman and Hall/CRC, Boca Raton (1994)

Clarke, B., Fokoue, E., Zhang, H.H.: Principles and Theory for Data Mining and Machine Learning. Springer, Heidelberg (2009)

Dudoit, S., van der Laan, M.: Multiple Testing Procedures with Applications to Genomics. Springer, Heidelberg (2007)

Efron, B.: Large-Scale Inference. Cambridge University Press, Cambridge (2010)

Kolaczyk, E.: Statistical Analysis of Network Data: Methods and Models. Springer, Heidelberg (2009)

Shmulevich, I., Dougherty, E.: Genomic Signal Processing. Princeton University Press, Princeton (2007)

Information Propagation in the Long-Term Behavior of Gene Regulatory Networks

Andre S. Ribeiro and Jason Lloyd-Price

Laboratory of Biosystem Dynamics, CSB Research Group
Dept. of Signal Processing, Tampere Univ. of Technology, Finland

Abstract. Gene regulatory networks (GRNs) constantly receive, process and send information. While stochastic in nature, GRNs respond differently to different inputs and similarly to identical inputs. Since cell types stably remain in restricted subsets of the possible states of the GRN, they are likely the dynamical attractors of the GRN. These attractors differ in which genes are active and in the amount of information propagating within the network. Using mutual information (I) as a measure of information propagation between genes in a GRN, modeled as finite-sized Random Boolean Networks (RBN), we study how the dynamical regime of the GRN affects I within attractors (I_A). The spectra of I_A of individual RBNs are found to be scattered and diverse, and distributions of I_A of ensembles are non-trivial and change shape with mean connectivity. Mean and diversity of I_A values maximize in the chaotic near-critical regime, whereas ordered near-critical networks are the best at retaining the distinctiveness of each attractor's I_A with noise. The results suggest that selection likely favors near-critical GRNs as these both maximize mean and diversity of I_A, and are the most robust to noise. We find similar I_A distributions in delayed stochastic models of GRNs. For a particular stochastic GRN, we show that both mean and variance of I_A have local maxima as its connectivity and noise levels are varied, suggesting that the conclusions for the Boolean network models may be generalizable to more realistic models of GRNs.

1 Introduction

Phenotypic diversity is critical for a species' survival [1]. One source of diversity in a monoclonal cell population is the stochasticity in gene expression [3, 17]. For example, *E. coli* selects for high noise in fluctuating environments [2]. In more complex organisms, stochasticity promotes phenotypic diversity within cells of the same cell type [18, 37]. More phenotypic diversity makes a cell population more likely to cope with unpredictable environmental fluctuations [2].

On the other hand, cells must behave reliably. In a pluri-cellular organism, cells of the same type must be robust, in that they need to respond similarly to similar external signals and behave similarly so as to reliably perform their functions. Thus, it is likely that their degree of phenotypic diversity is constrained, both within and between cell types.

Genes are embedded in Gene Regulatory Networks (GRN) [4], which constrains the dynamics of their expression levels. The topology of GRNs is likely to have evolved towards the maximization of the coordination between genes so as to robustly orchestrate a wide spectrum of behaviors [5], while maintaining responsiveness to environmental changes.

At the molecular dynamical level, it is unknown what a cell type is. However, even the simplest models of GRNs have a huge state space. Any given cell type must therefore be a very restricted subset of the states possible for the GRN. The simplest hypothesis is that cell types correspond to attractors of the GRN [6].

The first models of GRNs were Random Boolean Networks (RBNs) [6]. RBNs aim to capture the dynamics of GRNs arising from the interactions between genes. Nodes represent genes and can have two states: '1' if expressing and '0' otherwise. Nodes interact and update their state according to Boolean functions of the states of input genes. This model is an idealization of real GRNs and does not account for many features, but it captures, to some extent, how the topology constrains the dynamics of the genes' expression levels and the propagation of information between genes.

In its original formulation, RBNs are synchronous, deterministic systems [6]. Thus, they often have multiple attractors – state cycles into which the network settles – that have been associated to the cell types the GRN can express [6]. This hypothesis is problematic since GRNs are noisy, and therefore do not have deterministic attractors.

The concept of "cell type as attractor" was recently revisited and extended to explore its applicability to more realistic models of GRNs, namely, noisy RBNs and delayed stochastic GRNs, which account realistically for the noise at the molecular level as well for the time that complex processes such as transcription and translation take to occur [7].

In noisy RBNs, there is probability P of each gene "misbehaving" (to do the opposite of what its Boolean rule dictates at that time) at each time step. These "bit flips" perturb the network from its original trajectory, and place it in a new region of the state space. Generally, P is assumed to be small enough such that the system can return to a state cycle before the next perturbation occurs [7].

One way to quantify the effects of perturbations and the robustness of the attractors to this noise is to perturb each node of the network one at a time from every state of a given attractor [7], and observe the resulting trajectory of the system in the state space. For most perturbations, the RBN returns to the original attractor. However, there are usually several perturbations that cause the RBN to settle into another attractor. Attractors for which there is at least one perturbation that causes the system to leave the attractor were named "noisy attractors" [7].

It was also found that, in general, single bit flip perturbations are not sufficient to allow the system to go from any given attractor to every other attractor. Instead, there are usually sets of noisy attractors that can reach each other by single bit flips, but cannot reach other attractors. Such sets of noisy attractors, which the system can switch between by single bit flips, are named the "ergodic sets" of the system (following the standard definition of ergodic set in the context of stochastic processes). The ergodic set is such that fluctuations due to "internal noise" (here, single bit flips in the Boolean framework and stochastic fluctuations in the "delayed stochastic" framework) are not sufficient to make the system leave the ergodic set [7].

In other words, an ergodic set is a set of state cycles such that there is an undirected path between all the state cycles. A path exists between two state cycles when the system is able to reach the other state cycle due to genes' mistakes caused by internal noise in a state of one the state cycles.

It is noted that, in the Boolean framework, the number of ergodic sets depends on the level of noise (probability of a bit flip) while the number of noisy attractors does not (as these are identical to the attractors of the noiseless version of the Boolean network).

In general, noisy RBNs possess multiple ergodic sets, each composed of one or more "noisy attractors" [7]. Relevantly, multiple noisy attractors and ergodic sets also exist in the more realistic models of GRNs with delayed stochastic dynamics [7, 30].

RBNs have two distinct dynamical regimes, ordered and chaotic, separated by a phase transition dubbed "critical" [6]. The dynamical regime of a RBN is determined by its sensitivity, which in turn is determined by its mean connectivity (mean number of connections per node) and mean p_b (probability that the output of the Boolean transfer function is '1' for any set of input states) [31].

In "ordered" RBNs, the fraction of genes that remain dynamical after a transient period vanishes like $1/N$ as the system size N goes to infinity; almost all nodes become "frozen" on an output value (0 or 1) independent on the initial state of the network. In this regime, the system is strongly stable against transient perturbations of individual nodes. In "disordered" (or "chaotic") RBNs, the number of dynamical, or "unfrozen" nodes scales like N and the system is unstable to many transient perturbations.

In words, this means that chaotic RBNs tend to have wildly different responses to very similar inputs. In the biological setting, it would therefore not be advantageous for a cell to have a chaotic GRN, since in common environmental settings, similar inputs require similar responses. On the other hand, ordered GRNs respond identically to very distinct input signals, which in most situations would be disadvantageous as well. For that reason, near-critical GRNs are likely to be favored. If so, the topology and logic of evolved GRNs is likely to be constrained by its sensitivity. It is an attractive hypothesis that GRNs are near critical, since they would display a balance between robustness to random perturbations and flexible switching induced by targeted perturbations [7].

As stated, a typical attractor of a critical RBN is stable under most small, transient perturbations, but a few perturbations can cause a transition to a different attractor. This observation forms the conceptual basis for thinking of cell types as attractors of critical networks, since in complex organisms cells are both homeostatic in general and capable of differentiating when specific signals (perturbations) are delivered.

A property of networks is the propagation of information between its nodes. Because of this, in principle, it is possible to approximate the topology of the network from a trajectory of the network by using a measure of correlation such as mutual information [34, 35]. The propagation of information between the nodes in the network is related to its ability to exhibit the complex behaviors. This ability has been assessed by several methods in RBNs, and found to be maximum in the critical regime [9, 14, 11, 15, 16, 9].

A recent study assessed the amount of information propagated within RBNs when in attractors by calculating the temporal pairwise mutual information from time series of attractors (I_A). To have an overall assessment of the networks' capacity to propagate information, for each RBN, the I_A was averaged over many attractors. It was found that the mean I_A is maximized in the critical regime [5], implying that these networks are those that best propagate information between its nodes.

Strictly speaking, the I_A between the time series of two nodes of a network measures of how correlated, or coordinated in time, their behavior is. The correlations arise from the propagation of information between input and outputs, namely, of the input node's state at one moment in time, used to determine the output node's state at the next moment, according to the transfer function. It is in this sense that the I_A of a network in a given attractor can be used as a measure of the amount of information propagated between the nodes when on that attractor. Therefore, the mean I_A over many attractors has been used as global measure of the propagation of information in the network [5]. In this regard, critical networks were found to best propagate information between the nodes [5].

To the extent that the evolutionary fitness of a GRN depends on its ability to propagate information and enhance the coordination of behaviors between all genes, and RBN models capture the essential features of the organization of GRNs, critical networks are naturally favored. The maximization of mean I_A may therefore be a sensible proxy for maximization of fitness within an ensemble of evolutionarily accessible networks, as these networks can orchestrate the most complex, timed behaviors, possibly allowing robust performance of a wide spectrum of tasks. If so, the maximization of pairwise mutual information within the space of networks accessible via genome evolution may play an important role in the natural selection of real genetic networks [5].

The specialized, fully differentiated, cell types of pluri-cellular organisms differ in which set of genes is active, cell cycle length, etc., allowing them to perform distinct tasks. As such, they almost inevitably differ in (I_A). If selection shapes the structure and logic of GRNs regarding their ability to propagate information, then in pluri-cellular organisms, such selection ought to act at the cell type level. In other words, provided that noisy attractors are the possible long term behaviors of GRNs [7], it is likely to be selectively advantageous for each cell type to have maximum I_A, as long as they remain phenotypically (i.e, dynamically) distinct from cells of another type.

The diversity of the I_A of the various cell types is an expression the phenotypic diversity between and within the cell types. We hypothesize that, besides maximizing the mean I_A of all attractors, it is likely advantageous to maximize the number of distinct behaviors or tasks that can be performed by the multitude of cell types of an organism and by the cells of each cell type, while maintaining the distinctiveness of different cell types. Further, since the execution of cellular functions requires information propagation within the GRN [4], this diversity should be measurable in the diversity of values of I_A of the various attractors of the network. For this reason, we use the variance of the values of I_A of the attractors of the GRN as a measure of the diversity of behaviors or tasks that a GRN can perform.

So far, the distribution of I_A of attractors of RBNs (or other models of GRNs) has not been studied. We study the I_A of attractors in individual RBNs and the I_A distributions of ensembles of RBNs as a function of their global topological features, and thus the dynamical regime of the networks.

Since the dynamics of GRNs are noisy [3], it is important to evaluate the effects of noise on the I_A distributions. The strength of the molecular noise in RNA and protein levels has been shown to differ from gene to gene [21, 13] since it is sequence dependent, among other reasons [19]. It is therefore likely to be evolvable [20]. If so, evolving the noise strength may be a mechanism to evolve phenotypic diversity and thus the I_A distributions. We therefore study how introducing noise in the dynamics of RBNs affects the I_A distributions in each dynamical regime. Finally, we use a delayed stochastic GRNs [8], where the noise in gene expression and gene-gene interactions is more realistic, and show that the findings in RBNs have a parallel in delayed stochastic models of GRNs.

2 Methods

2.1 Boolean Networks

We generate RBN topologies imposing the number of nodes (N), the mean number of connections per node (k), and the probability that the output of the Boolean transfer function is '1' for any set of input states (p_b). RBNs with equal k and p_b are said to be of the same *ensemble* [12]. We generate topologies according to the "Random 2" algorithm from [22]. We only vary k, and always set the p_b to 0.5, as this is the most common method to generate RBNs in each of the dynamical regimes (ordered, critical, and chaotic). We model RBNs with $N = 250$ as the search for attractors (described below) becomes prohibitively expensive for larger N.

Noise in the dynamics of RBNs can be modeled in several ways. Here, we give each node a small probability η of assuming the opposite value from what its

Boolean rule specifies at each time step. Note that a bit flip corresponds to a large fluctuation in the protein level of a gene, and thus it should not be a very common event in real GRNs. Weaker fluctuations cannot be captured in this modeling strategy but it is noted that such small fluctuations commonly do not significantly affect the dynamics of the protein level of output genes [7].

2.2 Delayed Stochastic Models of Gene Networks

We follow the modeling strategy of delayed stochastic GRNs proposed in [8]. The models are implemented in the simulator SGNSim [28], and their dynamics are based on the "delayed SSA" [32], that, unlike the original Stochastic Simulation Algorithm [33], uses a waiting list to store delayed output events, proceeding as follows (t denotes time):

1. Set $t \leftarrow 0$, $t_{stop} \leftarrow$ stop time, read initial number of molecules and reactions, create empty waiting list L.
2. Do an SSA step for input events to get next reacting event R_1 and corresponding occurrence time t_1.
3. If $t_1 + t < t_{min}$ (the least time in L), set $t \leftarrow t + t_1$. Update number of molecules by performing R_1, adding delayed products into L as necessary.
4. If $t_1 + t \geq t_{min}$, set $t \leftarrow t_{min}$. Update number of molecules by releasing the first element in L.
5. If $t < t_{stop}$, go to step 2.

Multi-delayed reactions are represented as: $A \rightarrow B + C(\tau_1) + D(\tau_2)$. In this reaction, B is instantaneously produced and C and D are placed on a waitlist until they are released, after τ_1 and τ_2 seconds, respectively.

The modeling strategy of delayed stochastic gene networks [8] accounts for stochastic fluctuations and, regarding gene expression, by being a multiple-time delayed reaction, for the fact that transcription and translation are multi-step processes that take non-negligible time to complete once initiated. This modeling strategy was validated [26] by matching measurements of gene expression at the single RNA and protein level [25].

The delayed stochastic GRNs simulated here consist of 18 genes. For each gene i there is a multi-delayed, single-step, reaction for transcription and translation (reaction 1), and for protein degradation (reaction 2) (RNA polymerase is not explicitly modeled as they are assumed to be available in sufficient amounts). In these reactions, Pro stands for the promoter region of a gene and P_i for protein, while k_t and k_d are the stochastic rate constants of transcription initiation and protein degradation, respectively:

$$Pro_i \xrightarrow{k_t} Pro_i(\tau_1) + P_i(\tau_2) \qquad (1)$$

$$P_i \xrightarrow{k_d} \emptyset \qquad (2)$$

Each gene is assumed to have a promoter region with a transcription start site and a transcription factor binding site. For simplicity, all interactions are assumed to be repressive, since all genes have a basal expression level. Repression of Pro_i by P_j (expressed by Pro_j) is modeled by reaction 3 with rate k_r, while unbinding of the repressor occurs via reaction 4 with rate k_u, and its degradation while bound via reaction 5 with rate k_d (as in 2):

$$Pro_i + P_j \xrightarrow{k_r} Pro_i.P_j \qquad (3)$$

$$Pro_i.P_j \xrightarrow{k_u} Pro_i + P_j \qquad (4)$$

$$Pro_i.P_j \xrightarrow{k_d} Pro_i \qquad (5)$$

Unless stated otherwise, we set $k_t = 0.1\ s^{-1}$, $k_r = 100\ s^{-1}$, $k_u = 0.01\ s^{-1}$, and $k_d = 0.001\ s^{-1}$. Time delays are set to: $\tau_1 = 2s$ and $\tau_2 = 100s$. All values are within the range of realistic parameter values for *E. coli* [8, 26].

2.3 Measuring Mutual Information in Attractors

To estimate the flow of information in RBNs, in [5] I was computed from concatenated time series of multiple attractors. We follow a similar procedure, except that we do not concatenate the time series so that we measure the I_A of each attractor independently. I_A is calculated from the Shannon entropy of the time series of all pairs of nodes. For node i, its time series s_i is a binary string containing a 0 with probability $p_i(0)$ and a 1 with probability $p_i(1)$. The entropy of s_i is defined as: $H[s_i] = -p_i(0) \log_2 p_i(0) - p_i(1) \log_2 p_i(1)$.

Similarly, let $p_{ij}(xy)$ be the probability that nodes i and j are respectively in states x at time t and y at time $t+1$ where $x, y \in \{0, 1\}$. The joint entropy of the corresponding string of pairs s_{ij} is:

$$H[s_{ij}] = -\sum_{r,s=0}^{r,s=1} p_{ij}(rs) \log_2 (p_{ij}(rs)) \qquad (6)$$

The mutual information between two nodes is [5]:

$$I_{ij} = H[s_i] + H[s_j] - H[s_{ij}] \qquad (7)$$

I_{ij} measures the amount of information the state of node i at time t gives about the state of node j a moment later. The influence may be indirect through an ancestor node. The mean I of a RBN's time series is given by the average I_{ij} for all pairs.

We search for attractors by starting in a random state and updating the state until a state repeats. This search is costly in chaotic RBNs since transients can be very long, so we limit the search to 2^{13} steps as in [23]. The attractors not found by this search are likely to not be biologically relevant, as their length implies unrealistically long cell cycles. This search is performed starting from 5×10^4 random states to attempt to find almost all of the attractors in the network, as shown in [23]. Finally, the I_A of an attractor is calculated from the time series that includes the first state that repeats, and all subsequent states until the first repetition of that state (inclusively).

When the RBN's dynamics are noisy, the attractor search cannot proceed as described. Instead, the I_A is calculated by starting from a random state, running the network for 10^4 time steps (for large k, good convergence was obtained for discarded transients of length 10^3, as in [5]), and then measuring I from the following 10^4 states (for $k = 3$, the mean attractor length was found to be $\sim 10^3$).

In delayed stochastic GRNs [8], information propagates between genes via the proteins acting as transcription factors (TF), as in real GRNs. The information itself is, to some extent, in the amount of the TF, which affects the propensity with which the TF binds to the Transcription Factor Binding Site (TFBS), subsequently altering the transcription rate of the output gene. For simplicity, we do not model other mechanisms of expression regulation between genes, such as protein-protein interactions [24].

To measure I_A in delayed stochastic GRNs models, we proceed as follows. We design a GRN with multiple noisy attractors [7]. The GRN is initialized without proteins and simulated for a period of time which is long enough to reach one of the noisy attractors. From that moment onwards, the levels of all proteins are recorded in fixed intervals. All protein levels are then independently binarized using k-means as in [7], and I_A is computed from the binarized time series as in noisy RBNs.

Finally, we note that here we only measure mutual information between consecutive time moments and not between longer time lags. This is because the mutual information is measured only on attractors, after long transients, and thus already accounts, to some extent, for indirect interactions in the network [5]. Nevertheless it might be of interest in future studies to observe I_A for other time lags.

3 Results

3.1 Mutual Information Averaged over Many Attractors in Random Boolean Networks

We first revisit previous results. As mentioned in the introduction, a recent study assessed the information propagated within RBNs when in attractors, by calculating their I_A averaged over many attractors, for each RBN [5]. Fig. 1 shows the measured mean I_A, for networks of varying mean connectivity k. It maximizes near the critical regime. Note how the value of k for which the mean I_A maximizes approaches the critical value $k = 2$ as the system size grows.

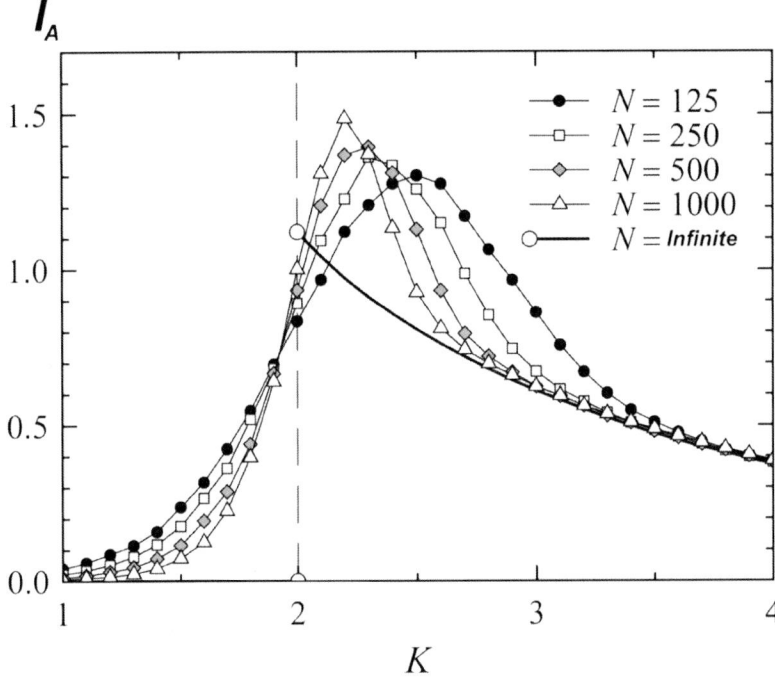

Fig. 1. I_A as a function of k for several different system sizes. For these calculation, $N = 10^4$, with 40 runs from different initial states per network. We discarded transient of length 10^4 updates for each run. (For large k, good convergence was obtained for discarded transients of length 10^3.) The sequences of states were recorded for a sample of $10N$ pairs of nodes in each network. The vertical dashed line indicates the critical value of k.

In accordance with these results, it was found that, in particular, the mutual information between subsequent states of individual nodes is maximized in the critical regime [10].

The value of mean connectivity for which mean I_A over all attractors maximizes changes with N because of the effects of local structures. These effects decrease with N, for any given k.

The local structure of a network can be characterized by the degree of clustering between the nodes [39]. The generalized clustering coefficient, C_p, accounts for all non-treelike local structures, such as self-connections, bidirectional connections, triangles, squares, etc. For a formal definition, refer to [39].

To observe how C_p affects I_A, the I_A was measured for RBNs with $k = 2$, removing various orders of C_p, averaging over many attractors. Fig. 2 shows that the higher the order of C_p fixed to zero, the less mean I_A varies with N [39]. Importantly, the effects of local topological features are not very significant for $N > 200$ [39]. For that reason, here we model networks with $N = 250$.

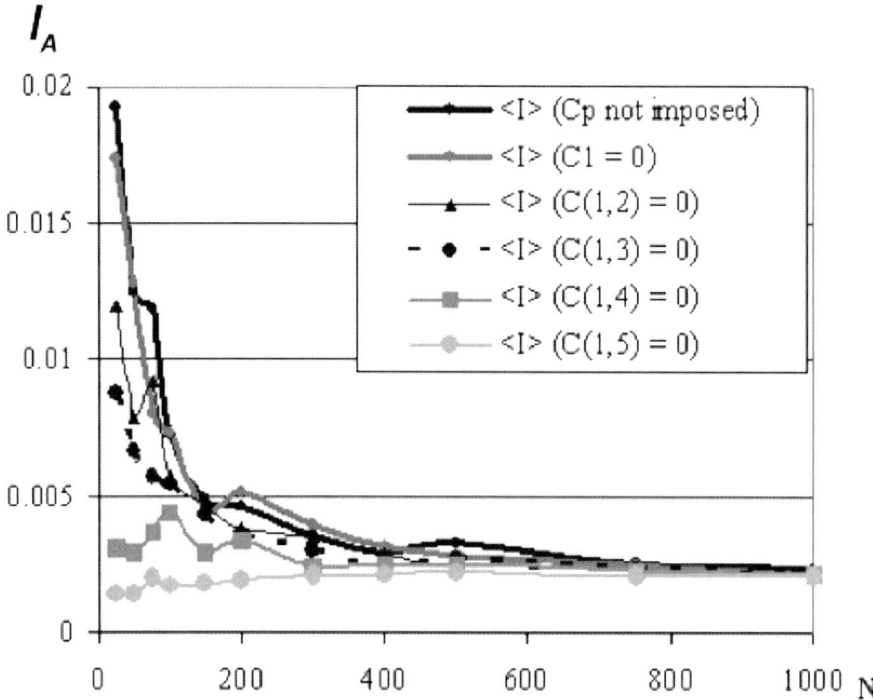

Fig. 2. I_A of RBNs computed from time series of length $T = 10^3$. The topologies are built such that up to various p the $C_p = 0$. Each data point is an average of 100 independent RBNs, varying N and with $k = 2$.

3.2 Random Boolean Networks

We first numerically assess the mean I_A values for varying k (Fig. 3, left). The maximization of mean I_A occurs in the chaotic near-critical regime at $k \approx 2.5$, due to finite size effects ($N = 250$ nodes). Previous results showed that I [5], calculated from concatenated time series of attractors, also maximizes at $k \approx 2.5$, when $N = 250$.

We now examine the distributions of I_A for individual RBNs. The attractors of a RBN differ in length and in the set of non-frozen nodes, among other things, causing their I_A to differ. Examples of such distributions for individual RBNs are shown in Fig. 3 (right) for an ordered, a critical and a chaotic RBN. These distributions are representative of the most common distributions found in each dynamical regime. Nevertheless, it is noted that these distributions vary wildly between RBNs with identical topological features. The peaks are at the I_A values of the various attractors of the three nets studied while the peak heights correspond to the fraction of times the search ended in an attractor with that I_A.

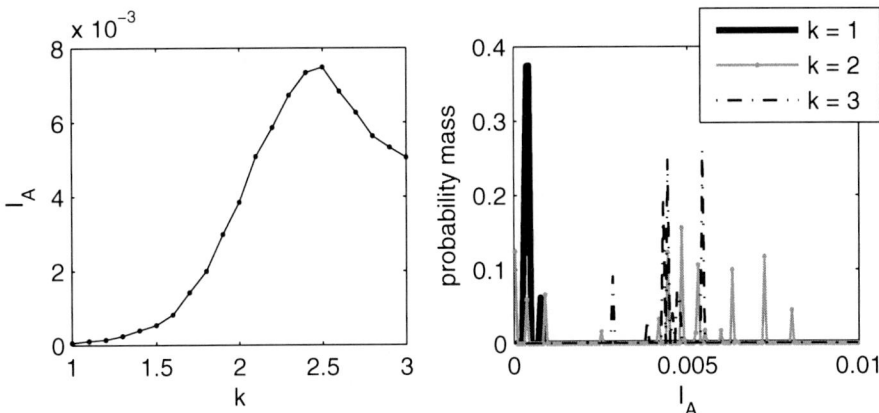

Fig. 3. Left: Mean temporal mutual information within attractors, I_A, as a function of the mean connectivity, k, 10^3 networks per point. Right: Sample I_A distributions of individual RBNs in different dynamical regimes. Number of nodes, N, equals 250.

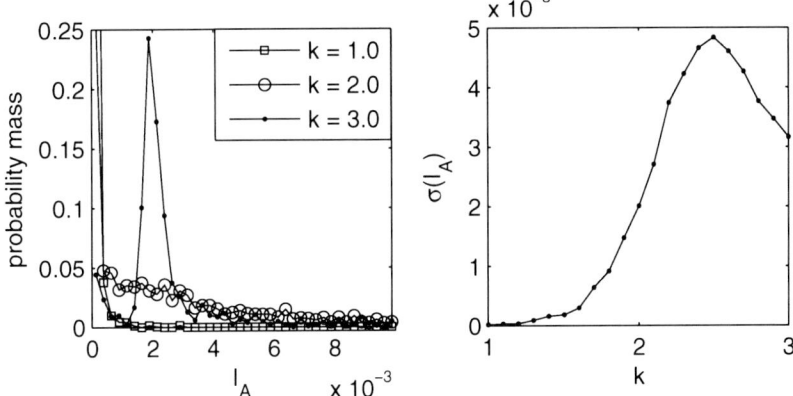

Fig. 4. Left: Distributions of temporal mutual information within attractors (I_A) from ensembles of RBNs with random topology and mean connectivity, k, equal to $1, 2, 3$. $N = 250$, 10^4 networks per distribution. Right: $\sigma(I_A)$ as k varies, 10^3 networks per point.

We generated ensembles of RBNs with varying k and plotted the overall distributions of I_A of individual attractors. Fig. 4 (left) shows the distributions for $k = 1, 2, 3$. The distribution significantly changes shape, although smoothly, with k (i.e. with the dynamical regime), from exponential-like for ordered and critical, to a bimodal distribution with a strong gaussian-like component for chaotic nets.

The gaussian-like component in chaotic RBNs is due to their lower number of frozen nodes and corresponding increase in I_A between unconnected nodes (i.e., spurious I_A). The non-triviality of the shape of the distributions indicates that the characterization of I_A of ensembles requires examining distributions rather than mean values alone. As the distributions of I_A vary widely between RBNs of the same ensemble, to characterize the behavior of the RBNs of an ensemble one needs to assess the I_A of many RBNs.

We computed the standard deviation ($\sigma(I_A)$) of I_A distributions as a simple means to compare the diversity of I_A values of RBNs of different dynamical regimes (Fig. 4, right). For finite sized RBNs, $\sigma(I_A)$ maximizes in the near-critical, slightly chaotic regime, for the same connectivity that mean I_A maximizes ($k = 2.5$).

We nevertheless note that, while in this case the value of $\sigma(I_A)$ is able to partially capture the effects of varying the mean connectivity k (namely, the increase in variability), because the distributions of I_A change shape with k, these need to be shown as well (Fig. 4, left), since $\sigma(I_A)$ alone is not sufficient to describe the change of shape of the distribution.

A note is needed regarding the effects of finite size on the connectivity for which both mean and diversity of I_A are maximized. First, this connectivity ($k = 2.5$) is the same for which I maximizes as well, in RBNs with $N = 250$ [5]. Since the value of k for which I maximizes approaches the critical value ($k = 2$) as one increases N [5], we hypothesize this to occur also for the mean and standard deviation of I_A. This needs to be confirmed in a future work in greater detail.

We note that we observed that both the mean and standard deviation of I_A maximized for slightly higher values of k in smaller networks ($N = 150$), as I does [5] (data not shown). This suggests that increasing N will cause the mean and diversity of I_A to maximize for values of k closer to the critical value, as I does [5]. Computational complexity limits the search for the attractors of networks significantly larger than $N = 250$.

Next, we study the I_A distributions of noisy RBNs. Previous studies have shown that for low noise levels (similar to the lower noise levels tested here), the network generally cycles 'around' the original attractors and rarely 'jumps' between attractors, thus have multiple ergodic sets [7]. We estimated that, introducing in RBNs the lowest noise level tested here ($\eta = 2 \times 10^{-4}$), generally causes the number of ergodic sets to decrease by roughly $\sim 25\%$, in agreement with the findings in [7]. This can also be seen indirectly from our measurements of I_A in noisy RBNs. As weak noise is introduced, each time the system cycles around a noisy attractor it yields a slightly different value of I_A. After many cycles (or many independent runs that lead to the same attractor), one ends up with a distribution of I_A values centered near the I_A of the noiseless attractor (Fig. 5a). If the system was jumping between attractors, one would expect the peaks of I_A to merge.

For the higher noise levels tested, the network does not remain near the original attractors, and the I_A of independent runs is always similar, resulting in a unimodal I_A distribution. While noise decreases the correlation between neighbors, it also temporarily unfreezes frozen nodes, leading to higher values of spurious I.

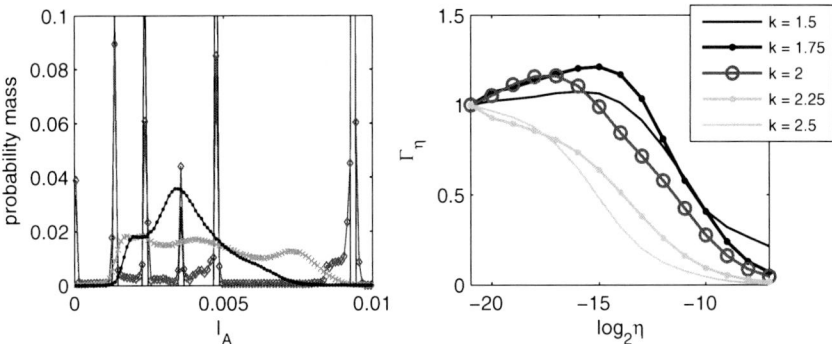

Fig. 5. Left: Distribution of temporal mutual information within attractors, I_A, from a single $k = 2$ network as noise is added to the dynamics. The noise strength, η, is set to 0 (solid black line), 4×10^{-5} (diamonds), 0.001 (gray x's), 0.002 (solid black circles). The graph has been scaled up to show detail. The network was started from 10^6 initial states for each distribution. Right: Relative change in standard deviation of I_A values, $\sigma(I_A)$, as a function of noise strength for RBNs with mean connectivity, k, equal to 1.5 (black line), 1.75 (black line with dots), 2 (circles), 2.25 (gray line with dots), and 2.5 (gray line). Size of the networks, N, is 250.

To exemplify how noise affects the I_A distributions of individual RBNs, in Fig. 5 (left) we show the I_A distribution of a critical RBN for various noise levels from 'low' to 'high'. By 'low noise' we imply that the number of bit flips due to noise is small enough so that the I_A values do not differ significantly from the noiseless case, while in the 'high noise' scenario they do. The strength of the noise is varied by changing the parameter η as previously described. We set the following values of η (probability of bit flipping for each node at each time step): 0, 4×10^{-5}, 0.001, and 0.002. Note that the distribution for $\eta = 2 \times 10^{-4}$ (low noise) is nearly identical to the one for $\eta = 0$, except that it is slightly broader around the original peaks, while for $\eta = 0.002$ (high noise) the distribution is unimodal.

Earlier, we found that in the absence of noise, chaotic near-critical RBNs maximize the mean and diversity of I_A values. Noise, at levels at which the original attractors are still retained, does not significantly affect the mean I_A. However, it does affect the diversity of I_A values. We measured this change in the degree of diversity of I_A values for increasing noise strength by the ratio between the standard deviation of I_A between each noisy case and the noiseless case ($\Gamma_\eta = \frac{\sigma(I_A(\eta))}{\sigma(I_A(0))}$). Results are shown in Fig. 5 (right).

In ordered ($k < 2$) and critical ($k = 2$) RBNs, as noise increases, the diversity $\sigma(I_A)$ initially increases (as the peaks broaden). Beyond a certain noise level, it starts decreasing as the behaviors within different attractors become indistinct (merged peaks). In chaotic networks ($k > 2$), even a small amount of noise decreases $\sigma(I_A)$ since small perturbations cause large avalanches of damage [15, 16], often causing the trajectory to leave the attractor. Ordered near-critical

RBNs ($k = 1.75$) have the highest relative increase in diversity as noise increases, and sustain this degree of diversity for very high noise levels (until the highest noise level tested). That is, these RBNs are more robust to noise in their ability to propagate information within the network.

It is noted that, for the highest noise levels tested, the notion of I_A loses some significance as the networks are unable to remain for a long time in any noisy attractor. As the noise level is increased, the number of ergodic sets decreases [7]. These highest noise levels were tested to verify how this loss of distinctiveness between ergodic sets affects I_A (mean and variance). As seen, it causes both of these quantities to decrease.

3.3 Delayed Stochastic Gene Networks

It is necessary to explore if the results in RBNs have any kind of parallel in more realistic models of GRNs. Delayed stochastic models of gene networks have been shown to realistically account for the noise in the chemical processes involved in gene expression [8], as well as for the duration of complex processes such the promoter open complex delay, which are known to affect the degree of stochasticity of transcription [27]. Relevantly, these models of GRNs have been shown to have noisy attractors and multiple ergodic sets [7, 30].

It is computationally prohibitive to simulate ensembles of delayed stochastic GRNs to test whether, at the ensemble level, these models of GRNs maximize mean I_A and diversity of I_A values for a given sensitivity, because the state space of parameter values and number of parameters is too large. In addition, the simulation of the dynamics if stochastic GRNs according to the delayed SSA is far more computationally complex [8]. Since we do not simulate ensembles of networks, the general conclusion that mean and variance of I_A maximizes in near-critical networks is not warranted for delayed stochastic GRNs.

However, it is possible to test whether the mean and diversity of I_A values of the noisy attractors from specific delayed stochastic GRNs [7] also vary with connectivity and noise strength as they do in RBNs ensembles. It is also possible to verify if there are values of noise and connectivity for which mean and diversity of I_A maximize for, in one case, a given mean connectivity, and in another, for a certain range of stochastic fluctuations in RNA and protein levels for specific GRNs.

If such local maxima of I_A exist for a specific network when its connectivity and noise are varied, then these global parameters should be subject to selection, as in RBN ensembles, since they affect information propagation within the network as well as the diversity of levels of propagation that delayed stochastic GRNs can realize when it is on its various noisy attractors.

The concept of topology and logic in delayed stochastic GRNs is more complex, as these features are determined by several parameters such as production and degradation rates of proteins and RNAs, binding and unbinding rate constants between TFBSs and TFs, etc. Here we opted to vary the noise strength in the GRN by varying the propensity of the interactions between TFs and TFBSs,

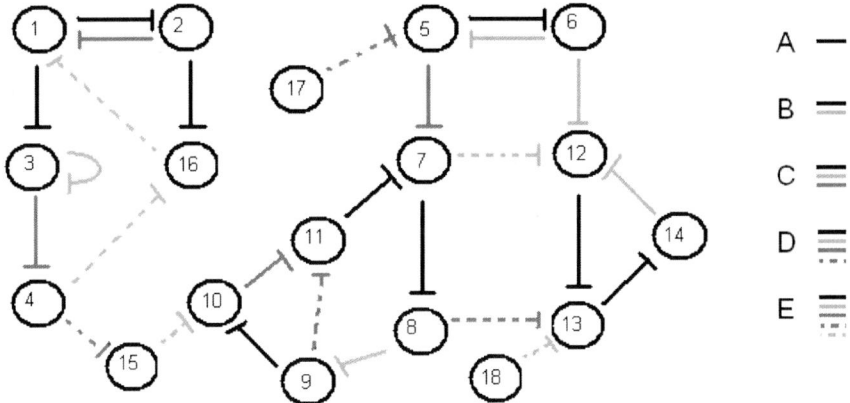

Fig. 6. Topology of effective connections of 5 stochastic GRNs with 18 genes and different number of connections. Net A connections are represented by the black lines, B are black and light gray lines, C are all solid lines, whereas D are all but the light gray dashed lines. Net E are all lines present.

while connectivity is varied by changing the number of TFs that, on average, can bind to each promoter region.

We implement delayed stochastic GRNs using reactions 1 to 5. The topology of five GRNs are depicted in Fig. 6, which differ in the number of connections. For clarity, only 'effective connections' are represented in the topologies. That is, only when a change in a protein level affects a target gene's expression level is that interaction depicted. Therefore, we expect that in the stochastic framework, I_A should maximize at approximately half the mean 'effective connectivity' for which it maximizes in the Boolean framework (in the Boolean framework, RBNs with $k = 2$ and $p_b = 0.5$ have a sensitivity of 1). Therefore, the mean connectivity in the stochastic framework is here denoted by "k_{eff}", as all connections are effective.

GRN 'A' has 9 effective connections, 'B' has 12, 'C' has 18, 'D' has 22 and, 'E' has 27. For these networks, k_{eff} is, respectively, equal to 0.5, 0.75, 1.0, 1.25, and 1.5, since $N = 18$ in the 5 networks. In the Boolean framework, these values would correspond, respectively, to mean k equal to 1.0, 1.5, 2.0, 2.5, and 3.0.

All the GRNs depicted in Fig. 6 have multiple noisy attractors [7]. These are mostly determined by the toggle switches (sets of two mutually repressing genes), each of which is a bistable circuit [29, 7]. Each switch has at any moment a probability of switching from one noisy attractor to the other that depends on the degree of fluctuations and mean protein levels of each gene [29]. The differences in the topologies A to E causes them to differ in number of noisy attractors. For example, network C has two bistable switches formed by the pairs of genes 1 and 2, and 5 and 6. After a transient, the expression levels of this set of genes can be in one of four "states", determining the noisy attractor [7]. Altering the topology of the switches alters the number of possible noisy attractors.

By inspection of the topology of these networks, and given previous studies of the dynamics of delayed stochastic toggle switches and repressilators (negative feedback loops with odd number of genes) [26, 29], it is possible to observe that the number of noisy attractors in each of the networks is less than 10, since the expression level of most other genes is determined by the 'noisy attractor' in which each of the switches is in. From our simulations, we observed that in all networks, all of its noisy attractors can be reached by initializing the system without any proteins (and all noisy attractors are equally likely), thus we opted to set this initial state in all simulations.

For each network, their noisy attractors differ in I_A, as some of the genes have different expression dynamics in different attractors. For example, in network C, when gene 5 is 'on' ('high' expression level) and gene 6 is 'off' ('low' expression level), the 5-gene repressilator (genes 7 to 11) will not oscillate periodically, while the 3-gene repressilator (genes 12 to 14) will.

We first tested the effects of varying the mean connectivity (k_{eff}). We simulated networks A to E, each 10^3 times, initialized with no proteins. Simulation time is 10^5 s, sampled every 10^3 s. We discard an initial transient of 10^4 s. In all networks $k_r = 100s^{-1}$.

Mean I_A values for GRNs A to E are show in Fig. 7(left). As seen, I_A changes significantly with k_{eff}, maximizing for $k_{eff} = 1$. It is thus possible to conclude that by varying the mean connectivity one can vary I_A of delayed stochastic GRNs and that there are values of connectivity that locally maximize mean I_A. The values of $\sigma(I_A)$ for varying k_{eff} are shown in Fig. 7(right). As seen, $\sigma(I_A)$ varies significantly with the mean effective connectivity, and there is a local maximum (for a slightly higher value of k_{eff} than for which the mean I_A maximizes).

We now test how the noise strength in the protein levels affects the distribution of I_A values in delayed stochastic GRNs. For this, we simulate networks with the same topology as network C (Fig. 6) but different values of k_r, which partially determines the strength of the noise in protein levels (other parameters could have been varied as well, such as the degradation rate of the proteins and/or RNA molecules). We therefore simulated 10^3 networks for each value of k_r, with the same protocol as in Fig. 7. The distributions of I_A when $k_r = 0.02, 0.2$ and $100s^{-1}$ are shown in Fig. 8, which is the stochastic equivalent of Fig. 5 (left). These distributions change shape with decreasing k_r (increasing noise), from bimodal to unimodal. Since the dynamics are stochastic, the I_A when on a noisy attractor varies between simulations. Thus, each noisy attractor is characterized by an interval of I_A values around a mean value (similar to noisy RBNs).

The similarity of the effects of varying the noise strength in this stochastic GRN and in the noisy RBNs is noticeable. In both cases, as the internal noise increases, the distribution of I_A values tends to change from multi-modal to uni-modal. That is, the noisy attractors lose their distinctiveness in I_A values as a consequence of their loss of 'stability' (the time that the GRN remains on a noisy attractor before 'jumping' into another noisy attractor diminishes with noise strength).

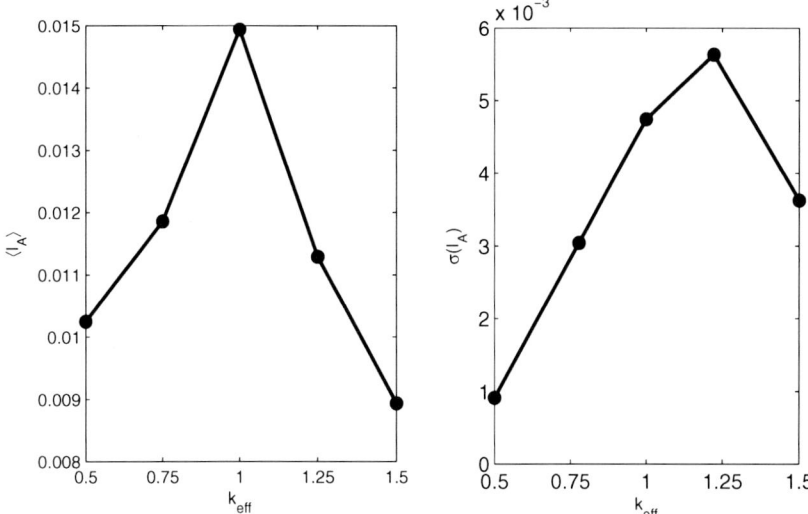

Fig. 7. Mean (left) and standard deviation (right) of temporal pairwise mutual information within attractors (I_A) of five delayed stochastic gene networks, differing in mean effective connectivity, k_{eff}, (topologies A to E).

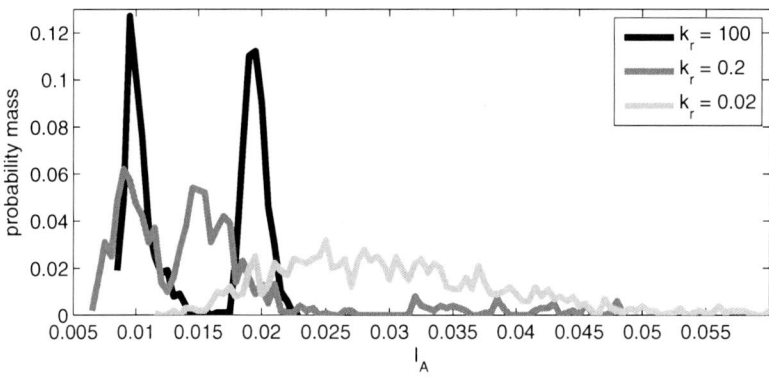

Fig. 8. Distribution of mean temporal mutual information within attractors, (I_A), in three delayed stochastic gene networks with topology C and different values for the rate of transcription repression, k_r.

As a consequence of this change in the shape of the distribution of I_A values, $\sigma(I_A)$ will vary significantly. $\sigma(I_A)$ is plotted in Fig. 9 for varying k_r between 10^{-2} and 10^6. It increases as k_r varies from 0.01 to $\approx 0.1 s^{-1}$, and decreases for further increases of k_r. In this regard, the $\sigma(I_A)$ of network C changes in a similar fasion to noisy RBNs, in that it maximizes for a given noise level. This is explained as follows. For very small k_r, repression is absent and thus each protein level only depends on rate of transcription k_t (reaction 1) and degradation k_d

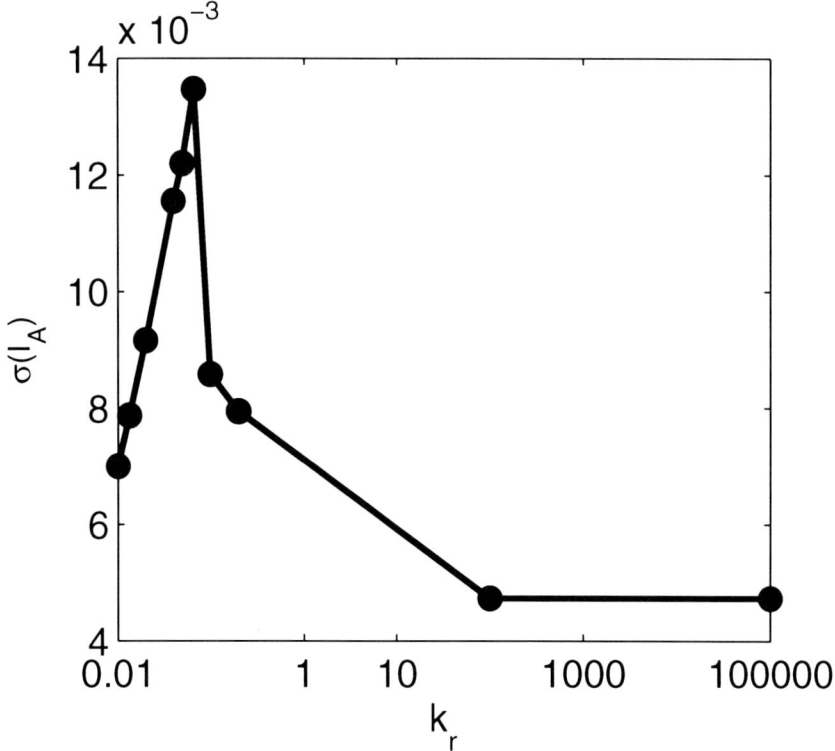

Fig. 9. Standard deviation of pairwise mutual information within attractors ($\sigma(I_A)$) in delayed stochastic GRNs with the topology of network C, differing in the rate of repression, k_r, and therefore in strength of the noise in protein levels.

(reaction 2), and will fluctuate around a mean value. The network will therefore not have much diversity in the values of I_A between independent simulations. On the other hand, very high k_r causes TFs and TFBSs to be too tightly bound, hampering dynamical changes. For example, the repressilator did not exhibit periodic dynamics within the simulation time. In this regime, I_A decreases with increasing k_r.

From these results, we find that these particular delayed stochastic GRNs, which differ in noise and connectivity, have different mean and variance of I_A. Both quantities were found to have local maxima.

The effects of varying k_{eff} are similar to what was observed in noiseless RBNs when varying the mean connectivity k. $\sigma(I_A)$ first increases with k_{eff}, then decreases for further increases, maximizing for a sensitivity around 1. The difference with regard to RBNs is that the mean connectivity for which $\sigma(I_A)$ maximizes is likely to depend on the value of k_r, among other parameters. However, we note that a parallel exists in RBNs in that the mean connectivity for which $\sigma(I_A)$ maximizes also varies, in that it depends on the p_b of the boolean functions in the RBN.

The effects of varying k_r (as a means to vary noise) on $\sigma(I_A)$ are also similar to the effects that varying the strength of the noise in noisy RBNs has on $\sigma(I_A)$. Namely, $\sigma(I_A)$ varies significantly with noise and there is a local maximum.

Even though we found a local maximum of I_A for a sensitivity of 1 when varying k_{eff}, one cannot, from these results, state that near-critical delayed stochastic GRNs maximize mean and variance of I_A of its noisy attractors. In fact, the concept of criticality has not yet been established for these models. Here, we extend this concept by adapting a common test for criticality in Boolean networks to delayed stochastic GRNs.

Specifically, one can do a perturbation analysis [15] as follows. A node's state is perturbed (in this case, change the protein level from 'low' to 'high', or vice versa) and then the trajectories of perturbed and non-perturbed networks are compared to quantify how many genes had their expression level affected after a certain time interval. A network is then defined to be critical if the number of perturbed genes is, on average, equal to 1.

We performed this test in networks A to E. After a transient of 10^5 s, we randomly chose one protein and altered its level. We then compared the levels of all proteins of the perturbed and non perturbed networks after an additional 5000 s. This was done 100 times per network, from which the mean number of perturbed genes was computed. From these tests, we verified that network C, with $k_{eff} = 1$, had the sensitivity closest to 1. From this, of these particular networks, the one that maximizes mean and variance of I_A is the one that is closest to being critical.

As mentioned, this result cannot be generalized to any ensemble of delayed stochastic GRNs, as the number of distinct parameters that can be used to vary connectivity, strength of connections, etc, is very large. Nevertheless, if the amount and diversity of information propagated between genes in various noisy attractors can be altered by changing the connectivity and noise in protein levels via the number of TFBS and their binding affinities (both of which are sequence dependent), then one can conclude that selection may in fact act upon the topology and logic of real GRNs so as to maximize these quantities. It is however not possible to conclude that there are are values of connectivity or noise strength that would maximize I_A and $\sigma(I_A)$ for an entire ensemble of stochastic GRNs. It also not possible to conclude from our results whether this maximization would occur in the near-critical regime for a given ensemble of stochastic GRNs.

4 Conclusions

By calculating I_A within individual attractors, we studied the distribution of I_A for ensembles of noiseless and noisy RBNs, as well as for a particular delayed stochastic GRN model as we varied noise and mean connectivity.

Regarding the ensembles of RBNs, we showed that in finite sized, noiseless RBNs, chaotic near-critical RBNs maximize the mean and diversity of I_A values of its attractors. Testing various levels of noise in the dynamics of noisy RBNs, we found that ordered near-critical RBNs best sustain the diversity of I_A in the face of internal noise in the dynamics of the RBNs.

Since the need to maximize mean and diversity of I_A favors networks in the chaotic near-critical regime, and the effects of internal noise on information propagation throughout the network are best coped with by ordered near-critical nets, we suggest that if these capacities are subject to selection, then they should cause GRNs to evolve structures which are dynamically between the chaotic near-critical and the ordered near-critical regimes. The precise dynamical regime of specific GRNs (ordered near-critical, critical, or chaotic near-critical) is likely to vary slightly from one organism to the next, depending on a multitude of factors, ranging from environmental to particular features of each GRN, such as its local topological features, etc.

Several features in RBNs are not realistic. For example, a central clock determines when genes' 'states' are updated [3]. Amongst the most detailed models of GRNs are the delayed stochastic models [8, 28], which account realistically for molecular kinetics as well as the time that complex processes such as transcription and translation take to occur, once initiated.

These delayed stochastic GRNs were found to have noisy attractors [7]. We used these models, to the extent that is currently computationally feasible, to partially verify the results obtained in RBNs. When measuring the I_A of the noisy attractors of the delayed stochastic GRNs, we found similar distributions of I_A to those observed in individual noisy RBNs. We also found dependency of mean I_A on the networks connectivity. Further, the dependence of $\sigma(I_A)$ on the noise strength and connectivity was also found, being similar to what was observed in RBNs.

It is sensible, as GRNs are parallel processing information systems, that their structure has evolved towards maximizing information propagation between its elements. Even delayed stochastic networks are found to preferentially remain in very confined regions of the state space, named "noisy attractors" [7], similar to what is observed in real cells [38]. These noisy attractors differ significantly on which genes and particular pathways in the GRN are active [38] and likely correspond to the various cell types in pluri-cellular organisms, each of which performs specialized tasks, and is phenotypically distinct of all others. If GRNs evolve towards maximizing the mean I_A of their attractors, then near-critical networks are favored.

In addition, complex organisms have evolved specialized cell types, to perform a wide range of tasks. We hypothesize that phenotypic diversity within cells of same type is maximized in pluri-cellular organisms, provided that cells of each type do not lose their phenotypic distinctiveness.

A GRN with multiple phenotypically distinct noisy attractors, where the phenotypic diversity within each noisy attractors is at maximum, will inevitably have a higher variance of I_A values than both a GRN with less noisy attractors and than a GRN with less diversity within each attractor, provided distinctiveness between noisy attractors. If so, our results suggest that near critical GRNs are favored, in that they maximize the variance of I_A without loss of cell type distinctiveness.

We do not imply that GRNs have purposely been selected to have a high diversity of I_A values, as it may have occurred for mean I_A. Instead, we suggest that the high variance of I_A values should emerge from the selection for high phenotypic diversity within and between specialized cell types.

Our findings at the level of ensembles of networks rely on the constraints in the dynamics of the networks of Boolean modeling strategy. However, it noted that GRNs of real organisms are known to possess the necessary mechanisms, on an evolutionary scale, by which they can regulate both the diversity between cell types as well as the diversity between cells of the same type. Namely, phenotypic distinctiveness between cells of different types depends on which genes and genetic pathways within the GRN are made active in each cell type. The number of distinct cell types is determined by the evolved differentiation pathways, which are known to be both externally driven as well as noise-driven [18]. Finally, cells of the same type can regulate the phenotypic diversity among them by regulating the degree of stochasticity of the genes active in that cell type [36, 37]. Provided these mechanisms, we suggest that both mean and variance of I_A may be subject to natural selection.

Phenotypic variation is critical since it is a necessary condition for evolution to occur [1]. This principle is likely to apply not only to the variability between organisms within a species, but also to the variability between cell types within an organism, and even the variability between cells within a cell type.

If the fitness of GRNs depends on the ability to propagate information reliably between the genes, and if the fitness of organisms depends on cell-to-cell phenotypic diversity both within and between cell types, our results suggest that near-critical GRNs are naturally favored within typical ensembles of GRNs' topologies.

A final note is necessary regarding the use of ensembles to study GRNs. Namely, the topology of real GRNs is not random, as these have evolved specialized topologies over millions of years. In that sense they may not be part of any ensemble. Besides, critical networks are not necessarily the best at propagating information of all possible topologies. Our results only show that they are best, *in general*, in comparison with ensembles of ordered and chaotic random networks. What appears certain is that real GRNs are highly efficient in propagating information between its elements, as cells are capable of accurately carrying out complex, temporally ordered and constrained processes, in a repeatable fashion.

For the above reasons, rather than evolving to be critical, we find it more plausible that GRNs have been selected to be reliable parallel information processing systems, and that selection occurs at the level of each cell type realized by the GRN. If so, we hypothesize that GRNs may have evolved to be critical, at least at the level of small gene clusters within the GRN, especially if these clusters receive and process information from external signals, since chaotic gene clusters would produce unreliable responses and ordered ones would be unable to respond differently to different inputs.

References

[1] Mayr, E.: Foreword in Variation. In: Hallgrimsson, B., Hall, B.K. (eds.) Variation. A Central Concept in Biology, Elsevier Academic Press, Amsterdam (2005)
[2] Acar, M., Mettetal, J., van Oudenaarden, A.: Stochastic switching as a survival strategy in fluctuating environments. Nature Genetics 40, 471–475 (2008)
[3] Kaern, M., Elston, T., Blake, W., Collins, J.J.: Stochasticity in gene expression: from theories to phenotypes. Nat. Rev. Genet. 6, 451–464 (2005)
[4] Mayr, E.: What evolution is. Basic Books, New York (2001)
[5] Ribeiro, A.S., Kauffman, S.A., Lloyd-Price, J., Samuelsson, B., Socolar, J.: Mutual Information in Random Boolean models of regulatory networks. Phys. Rev. E 77, 011901 (2008)
[6] Kauffman, S.A.: Metabolic Stability and Epigenesis in Randomly Constructed Genetic Nets. Journal of Theoretical Biology 22, 437–467 (1969)
[7] Ribeiro, A.S., Kauffman, S.A.: Noisy attractors and ergodic sets in models of gene regulatory networks. J. Theor. Biol. 755, 743–755 (2007)
[8] Ribeiro, A.S., Zhu, R., Kauffman, S.A.: A General Modeling Strategy for Gene Regulatory Networks with Stochastic Dynamics. Journal of Computational Biology 13(9), 1630–1639 (2006)
[9] Krawitz, P., Shmulevich, I.: Basin Entropy in Boolean Network Ensembles. Phys. Rev. Lett. 98, 0158701 (2007)
[10] Luque, B., Ferrera, A.: Measuring mutual information in random boolean networks. Complex Syst. 12, 241–252 (2000)
[11] Bertschinger, N., Natschlager, T.: Real-Time Computation at the Edge of Chaos in Recurrent Neural Networks. Neural Comput 16(7), 1413–1436 (2004)
[12] Kauffman, S.A.: The Ensemble Approach to Understand Genetic Regulatory Networks. Physica A 340, 733–740 (2004)
[13] Fraser, H., Hirsh, A., Giaever, G., Kumm, J., Eisen, M.: Noise Minimization in Eukaryotic Gene Expression. PLOS Biology 2(6), 834–838 (2004)
[14] Nykter, M., Price, N., Aho, T., Kauffman, S.A., Yli-Harja, O., Shmulevich, I.: Critical networks exhibit maximal information diversity in structure-dynamics relationships. Phys. Rev. Lett. 100, 058702 (2008)
[15] Rämö, P., Kesseli, J., Yli-Harja, O.: Perturbation avalanches and criticality in gene regulatory networks. J. of Theor. Bio. 242(1), 164–170 (2006)
[16] Serra, R., Villani, M., Semeria, A.: Genetic network models and statistical properties of gene expression data in knock-out experiments. J. of Theor. Bio. 227, 149–157 (2004)
[17] Suel, G.M., Garcia-Ojalvo, J., Liberman, L.M., Elowitz, M.B.: An excitable gene regulatory circuit induces transient cellular differentiation. Nature 440(23), 545–550 (2006)
[18] Chang, H.H., Hemberg, M., Barahona, M., Ingber, D.E., Huang, S.: Transcriptome-wide noise controls lineage choice in mammalian progenitor cells. Nature 453(7194), 544–547 (2008)
[19] Blake, W.J., Balazsi, G., Kohanski, M.A., Isaacs, F.J., Murphy, K.F., Kuang, Y., Cantor, C.R., Walt, D.R., Collins, J.J.: Phenotypic Consequences of Promoter-Mediated Transcriptional Noise. Molecular Cell 24, 853–865 (2006)
[20] Maamar, D., Raj, A., Dubnau, D.: Noise in gene expression determines cell fate in *Bacillus subtilis*. Science 317, 526–529 (2007)
[21] Barkai, N., Leibler, S.: Circadian clocks limited by noise. Nature 403(6767), 267–268 (2000)

[22] Airoldi, E., Carley, K.: Sampling Algorithms for Pure Network Topologies. Sig. Exp. Newsl. 7(2), 13–22 (2005)
[23] Samuelsson, B., Troein, C.: Superpolynomial Growth in the Number of Attractors in Kauffman Networks. Phys. Rev. Lett. 90, 098701 (2003)
[24] Holmberg, C.I., Tran, S.E.F., Eriksson, J.E., Sistonen, L.: Multisite phosphorylation provides sophisticated regulation of transcription factors. Trends in Biochemical Sciences 27, 619–627 (2002)
[25] Yu, J., Xiao, J., Ren, X., Lao, K., Xie, S.: Probing Gene Expression in Live Cells, One Protein Molecule at a Time. Science 311, 1600–1603 (2006)
[26] Zhu, R., Ribeiro, A.S., Salahub, D., Kauffman, S.A.: Studying genetic regulatory networks at the molecular level: Delayed reaction stochastic models. J. of Theo. Biol. 246, 725–745 (2007)
[27] Ribeiro, A.S., Häkkinen, A., Mannerstrom, H., Lloyd-Price, J., Yli-Harja, O.: Effects of the promoter open complex formation on gene expression dynamics. Phys. Rev. E 81(1), 11912 (2010)
[28] Ribeiro, A.S., Lloyd-Price, J.: SGN Sim, a Stochastic Genetic Networks Simulator. Bioinformatics 23, 777–779 (2007)
[29] Ribeiro, A.S.: Dynamics of a two-dimensional model of cell tissues with coupled stochastic gene networks. Phys. Rev. E 76, 051915 (2007)
[30] Dai, X., Yli-Harja, O., Ribeiro, A.S.: Determining the noisy attractors of a delayed stochastic Toggle Switch from multiple data sources. Bioinformatics 25(18), 2362–2368 (2009)
[31] Shmulevich, I., Kauffman, S.A.: Activities and Sensitivities in Boolean Network Models. Phys. Rev. Lett. 93(4), 48701 (2004)
[32] Roussel, M.R., Zhu, R.: Validation of an algorithm for delay stochastic simulation of transcription and translation in prokaryotic gene expression. Phys. Biol. 3, 274–284 (2006)
[33] Gillespie, D.T.: Exact Stochastic Simulation of Coupled Chemical Reactions. The Journal of Physical Chemistry 81(25), 2340–2361 (1977)
[34] Hecker, M., Lambecka, S., Toepferb, S., van Somerenc, E., Guthkea, R.: Gene regulatory network inference: data integration in dynamic models-a review. Biosystems 96, 86–103 (2009)
[35] Kaleta, C., Goehler, A., Schuster, S., Jahreis, K., Guthke, R., Nikolajewa, S.: Integrative inference of gene-regulatory networks in *Escherichia coli* using information theoretic concepts and sequence Analysis. BMC Systems Biology 4(116) (2010)
[36] Samoilov, M., Price, G., Arkin, A.P.: From Fluctuations to Phenotypes: The Physiology of Noise. Sci. STKE 366 (2006)
[37] Wernet, M., Mazzoni, E., Celik, A., Duncan, D., Duncan, I., Desplan, C.: Stochastic spineless expression creates the retinal mosaic for colour vision. Nature 440, 174–180 (2006)
[38] Margolin, A., Nemenman, C., Basso, K., Klein, U., Wiggins, C., Stolovitzky, G., Favera, R., Califano, A.: ARACNE: an algorithm for the reconstruction of gene regulatory networks in a mammalian cellular context. BMC Bioinformatics (suppl. 1) (2006)
[39] Ribeiro, A.S., Lloyd-Price, J., Kesseli, J., Häkkinen, A., Yli-Harja, O.: Quantifying Local Structure Effects in Network Dynamics. Phys. Rev. E 78(5), 056108 (2008)

Natural Language and Biological Information Processing

Samuli Niiranen, Jari Yli-Hietanen, and Olli Yli-Harja

Computational Systems Biology Research Group, Department of Signal Processing, Tampere University of Technology, Tampere, Finland

Abstract. Our ability to use natural language to communicate and cooperate with others is one of the defining characteristics of human intelligence. Much work has been put into developing theories which would explain the structure of language and how it relates to the information processing capabilities of the human mind. In this chapter we philosophically discuss natural language faculty as a being a mechanism conveying embodied information and natural language as such information.

1 Introduction

What is common to humans, ants, bees, birds and even bacteria? They all communicate or, in other words, exchange information with each other. Apart from bacteria, all these life forms are capable of symbolic communication. For example, bees communicate with each other to convey the direction and distance of a nutrition source via an intricate dance. Singing birds also learn regular tweets and other vocalized sounds [1] to indicate territorial boundaries and to attract mates. Animal communication serves the evolutionary purpose of facilitating the coordination of co-operation among groups of animals sharing the same form of communication, the same language. Thus, it usually involves things in the immediate local environment and other practical matters helping an animal to survive. Warning shrieks, territorial behavior, pack pecking order announcements, threats and expressions related to mating and nutrition are all examples of such communication.

Can these forms of animal communication be called language? Usually, the word language is reserved for the various human languages. The defining characteristics of a language are the symbolic nature of its expressions, the connection between a symbol and its intent, and a grammar which makes language a fundamentally open system of communication.

More specifically, Noam Chomsky, one of the pioneers of formal linguistics, and colleagues [2] have argued that human language faculty is unique in its capability to process hierarchically structured sequences of communication. Hauser et al [3] have shown that while non-human primates can learn and process sequences determined by local transition probabilities, they are unable to do so for more complex hierarchical structures with recursive embeddings. One way [4] to illustrate the distinction between these two sequence types is to look at them as being generated by two different formal grammar types: Finite-State Grammars (FSG) and Phrase Structure Grammars (PSG). The former is fully determined by local transitional probabilities between a finite number of items while the latter enables the generation of phrase structures via recursive rules. As shown in Fig. 1, an FSG is able to generate sequences such as ABAB or ABABAB with a two letter alphabet of A and B while a PSG can generate sequences like AABB where the [AB] part is embedded as in A[AB]B.

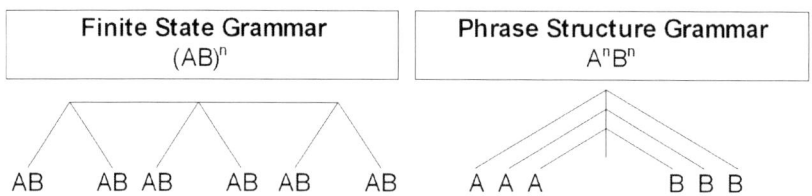

Fig. 1. Structure of an FSG grammar and a PSG grammar[4]

In fact, it has been experimentally verified [4] that different areas of the human brain process sequences generated with a PSG than with an FSG. The brain regions for recursive processing are also phylogenetically younger which supports the idea that such capability is a late evolutionary feature possibly existing only in the human central nervous system. However, some scholars, e.g. Steven Pinker [5], consider the ability to process recursive structures as only one of many crucial aspects of human language faculty differentiating it from those of animals. Further undermining the argument, a songbird species, the European starling, has been reported to accurately recognize acoustic patterns with recursive and hierarchical structures [6]. Still, it is clear that human natural language is by far the most complex and diverse of the forms of communication employed by any animal. The exact features, and their evolutionary history, which distinguish human languages from the forms of communication used by animals is an open research question.

In addition to recursion, other proposed unique features of human intelligence related to our language faculty include [7]:

1. voluntary control of the voice, face and hands
2. ability to imitate observed motor action
3. human-like teaching where the teacher observes the novice

All human languages share many structural properties of which recursive structure is just one, albeit very general and universal, example. A key challenge of cognitive science and linguistics is to give an explanation for the structure of language. Figure 2 illustrates a model for the structure of language arising from the interaction of three complex adaptive systems: learning, evolution and culture.

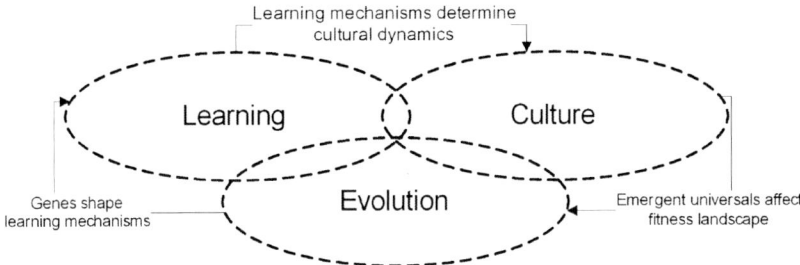

Fig. 2. The structure of language arising from the interaction of three complex adaptive systems[8]

Our biological endowment, our genes, provides us with the basic learning mechanisms to acquire language. This machinery is the mechanism through which language is transmitted culturally over time in human populations. Moreover, this process of cultural transmission leads to a set of language universals constrained also by our innate linguistic predispositions. Finally, the structural characteristics emerging from this process will affect the fitness of individuals using these languages leading in turn to the evolution of language learners, closing the loop of three interacting systems. The key thing to note here is the interplay between genetic predispositions, cultural transmission and biological evolution.[8]

Figure 3 illustrates the link between genetic predispositions and language structure. Genes provide the basis for mechanisms for learning and processing language. They determine our innate predispositions in the context of language acquisition. A predisposition is a property of an individual, but the final structure of natural language emerges from interactions in a population of individuals over time. Thus, cultural transmission bridges the link between predispositions and universals. While genes code the predispositions, biological fitness is partially governed by the extended phonotype (i.e., language structure).

A key question here is how strong the innate constraints on language learning are. Kirby et al. [8] argue that cultural transmission can magnify weak predispositions into strong linguistic universals undermining the hypothesis which assumes strong innate constraints.

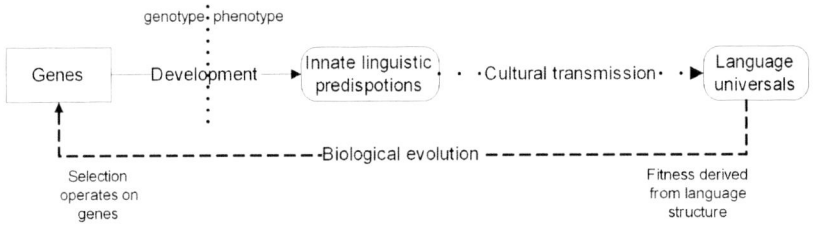

Fig. 3. The connection between biological predispositions and language structure[8]

In summa, human language faculty is a part of the biological information processing system known as the human cognitive system. It serves a key function as its system of symbolic communication. The extent to which and how human language differs from animal systems of communication remains an open research question. In any case, our language faculty arises from the interactions of three complex systems (learning, evolution and culture). Again, the extent to which each of these systems constrains the structure of human language is a matter subject to continuing debate.

We will next look a hierarchy of generic information processing systems and see how living beings with language faculty are situated in this context.

2   A Hierarchy of Information Processing Systems

Thinking about how artificial systems and living organisms relate to their environment and process information we can differentiate five distinct classes. The key separating characteristics are the level of adaptability to changes in operation environments and the level of autonomy the system exhibits.

2.1   Context-Free Closed System

In the first class, a context-free closed system, all inputs of the system are equal in the sense that the different input states carry no meaning on how the inputs are processed by the system. An example of such a system is one providing lossless compression of the input using the Lempel-Ziv algorithm. Such lossless compression functions by searching repeated symbol sequences in the system input and by replacing these with references to a single instance of the repeated sequence.

2.2   Context-Sensitive Closed System

The second class, a context-sensitive closed system, has as one input a static, context-specific model that contains information on how the other inputs are processed by the system. An example of a system in this class is a lossy audio compression system. Such a system can provide increased compression efficiency over lossless compression by omitting signal components not meaningfully audible to the human hearing system. In other words, the way in which the signal is compressed is sensitive to the static characteristics of the operation environment.

2.3   Open System with Feedback

In the third class, an open system with feedback, a relation exists between system output states and its input states. In such a feedback control system, the system partially captures the dynamic characteristics of the operation environment as input states are coupled to output states. An example of such a system is a temperature controller. This type of system is able to adapt to diverse changes in the operation environment (e.g., the temperate controller is able to maintain a constant temperature in an apartment when the outdoors temperature changes or someone leaves a window open). In contrast to the context-sensitive closed system, the system is able to monitor changes in the operation environment and adapt to them.

2.4 Open System with Feedback and Energy Management

The fourth class, an open system with feedback and autonomous energy management, extends the previous class by including autonomous energy management as one parameter in feedback control. In other words, not only are the input and output states coupled but the system has finite resources to carry out its operation of mapping input states to output states. The relation of different inputs and outputs states to internal energy consumption creates additional information. An example of such as system is an autonomous robot.

2.5 Open System with Feedback, Energy Management and Reproduction

The fifth system class, an open system with feedback, autonomous energy management and reproduction, further extends the previous class by being reproductive. As such, these systems are subject to competition and natural selection in an environment with limited resources which the systems utilize to operate and reproduce. The systems adapt to specific environments through successive reproductive generations as guided by exhibited fitness related to survival and successful reproduction. This selective process creates new information preserved by the accumulation of fitness-enhancing functionality over system generations.

Living organisms belong to this general class of systems and exhibit adaptation to the environment not just through genotypic differentiation over generations but also through learning supported by intra-species communication during their lifetimes. More specifically, living organisms are able to adapt to novel changes in the environment as their genetic endowment provides for behavioral degrees of freedom. As discussed in the introduction, a key component of this adaptive endowment is the use of natural language as a communication tool. In vernacular use of language, natural language usually refers only to the various human languages. However, the means of communication utilized by other species share many characteristics with human language. Audible cues are used by wolf packs to guide the path of the lead wolf in a hunting situation and we are all familiar with the warning shrieks animals use to warn others about the presence of a predator. The common thread in language is that it is used to coordinate the co-operation of individuals usually belonging to a single species, or in the case of humans, a single language group. Of course, human language is much more complex than the languages used by animals by including such elements as compositionality, recursion and abstraction as discussed in the introduction. Still, it is fundamentally a tool to facilitate human co-operation.

Looking further at the general characteristics of human language, it includes redundant information for error correction purposes as it is communicated over a noisy channel. A typical syntactic feature serving such a redundancy function is agreement. For example, in English there is an agreement between the grammatical person of the verb and the grammatical number of the subject. This is morphologically evident in the singular third person where the verb is usually inflected with the suffix -s ('I see', 'You see', 'He see-s'). It is easy to imagine why human language developed such redundant features. The environment in which a hunting party has to co-operate is sure not to be free of noise. To put things simply, a hunter listening to his colleague can gain additional information on whether the other party is communicating something about a third person through verbal agreement even if he doesnt capture the full signal (' plans to go in towards that that tree' vs. ' plan to go in towards that tree').

Claude Shannons seminal mathematical theory of communication formalizes the problem of transmission of information over a noisy channel in the context of digital data communication. However, as noted by Shannon, information theory does not deal with the semantic aspects of communication [9]. Redundancy in natural language cannot be separated from semantics, i.e. features which contribute most to the successful transport of critical meaning, in the context of typical human co-operation, are often those made redundant. The aforementioned person-number agreement is an example of semantically motivated redundancy in natural language. Also, communicating people often add redundancy to those parts of the message with most pragmatic significance in that specific situation of co-operation.

In general, it can be argued that information theoretical or other formal methods are inadequate for the study of natural language. If understood in the context of a closed system, natural language is not formally computable. Looking back at the presented hierarchy of information processing systems, this should not come as a surprise. Human language is adaptation for communication and information management in a fundamentally open system, our daily environment. However, although we claim that language is not algorithmic, it is also evident that it has structure. In contrast to formal languages, the structure of natural language is non-propositional in the sense that it does not make formal assertions about the world or convey truth values in the context of a specific utterance. It does exhibit a non-propositional structure mirroring our environment and us. Natural language can be said to be fundamentally embodied in both these contexts.

3 Natural Language and Embodiment

3.1 Embodied Cognition

The school of cognitive science which states the characteristics of the human mind are primarily determined by the form of the human body is entitled embodied cognition. This embodied mind thesis stands in opposition to other theories of the mind, namely those considering the mind an algorithmic information processing system such as cognitivism and computationalism.

Margaret Wilson has presented a list of six views which characterize embodied cognition [10]:

1. Cognition is situated. Cognitive activity happens in the context of a real-world environment and inherently involves perception and action. One example of this is moving around a room while trying to decide where the furniture should go at the same time. Another example of situated cognition

is day-dreaming. One is in a situation, but not in the situation or 'present'. During day-dreaming you may be doing something, but your mind or your thoughts are somewhere else.
2. Cognition is time-pressured. Cognition should be understood in terms of how it functions under the pressure of real-time interaction with the environment. When one is under pressure to make a decision, the choice made emerges from the confluence of pressures that one is under while in their absence a decision may be completely different. Since there was pressure, the result was the decision you made.
3. Off-loading cognitive work onto the environment. Due to limits of our information-processing abilities, we exploit the environment to reduce our cognitive workload. We make the environment hold or sometimes even manipulate information for us, and we harvest that information only on a need-to-know basis. This is evident in the fact that people use calendars, PDAs and other means to help them with everyday functions. We write things down so that we can use it when we need it instead of taking the time to memorize it.
4. The environment is part of the cognitive system. The flow of information between the mind and the world is so dense and continuous that the mind alone is not a meaningful unit of analysis. This means that the production of cognitive activity is not internal to the mind, but rather is a mixture of the mind and the environmental situation that we are in. Thinking and decision making are impacted by the environmental situations we operate in and is thus contextual.
5. Cognition is for action. The fundamental function of the mind is to guide action and perception as well as memory must be understood in terms of their contribution to situation-appropriate behavior. This claim has to do with the visual and memory perception of the human mind. Our vision is encoded into our minds as a what and where concept alluding to the structure and placement of an object. Our perception of what we see comes from our experience and exposure of it. Memory in this case does not necessarily mean memorizing something as such. Rather remembering a relevant point of view instead of as it really was in an objective sense.
6. Off-line cognition is body-based. Even when disconnected from the environment, the activity of the mind is grounded in mechanisms that evolved for interaction with the environment: sensory processing and motor control. This is best shown with infants or toddlers. Children use skills they were born with, such as sucking, touching, and listening, to learn more about the environment. The skills are broken down into five sensorimotor functions:

   (a) Mental imagery means visualizing something based on your perception of it.
   (b) Working memory short-term memory
   (c) Episodic memory long-term memory
   (d) Implicit memory means by which we learn certain skills until they become automatic for us.

(e) Reasoning and problem-solving having a mental model of something will increase the number of problem-solving approaches.

3.2 Cognitive Linguistics

Building on this philosophical foundation, George Lakoff and colleagues have studied embodied cognition in the context of linguistics. They argue [11] that the embodiment hypothesis of embodied cognition entails that our conceptual

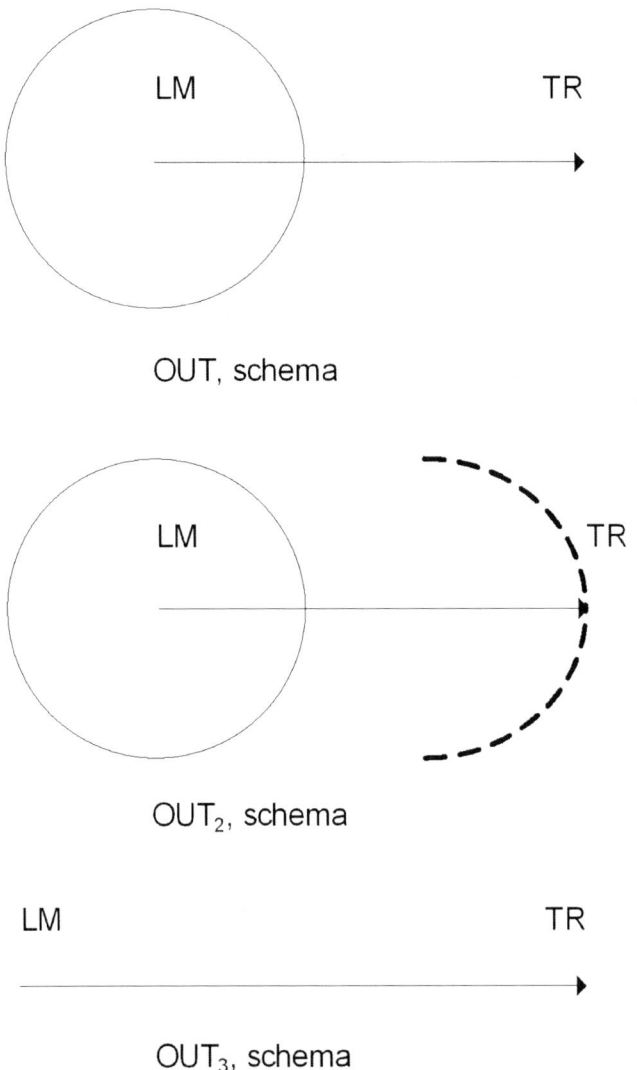

Fig. 4. Containment image schema (applied to the English word 'out')[13]

structure and linguistic structures are formed and shaped by the characteristics of our perceptual structures. As evidence, they cite research on embodiment effects from mental rotation and mental imagery, image schemas, gesture, sign language, color terms, and conceptual metaphors among other examples. Succinctly, cognitive linguistics [12] says that language is both embodied and situated in a specific environment.

As an example of such linguistic structuring, figure 4 shows how the containment image schema can be applied to the English word out in three different ways reflecting the various spatial senses of the word.

First, in the most prototypical case, 'out' is utilized where a clearly defined trajector (TR) leaves a landmark bounded spatially (LM), a clearly defined container: "Scott got out of the train" (top of figure 4). Second, 'out can also be used to indicate situations where the trajector is a mass spreading out, expanding the area of the containing landmark: "He poured out the beads" (middle of figure 4). Third, 'out' can also be utilized to describe motion of a trajector along a linear path where the containing landmark is not specified but implied: "The ship started out for Southampton" (bottom of figure 4). Most importantly, the natively spatial image schemas are used metaphorically to lend logic to non-spatial situations. For example, the in the sentence "George finally came out of his depression" a spatial containment schema is projected metaphorically onto emotional life.

Expanding on this, cognitive linguists argue that an embodied philosophy of mind would show the laws of thought to be metaphorical in general, not logical. Truth would be a metaphorical construction, not an attribute of objective reality. That is, it would not rely on any foundation ontology as might be sought in the physical sciences or religion, but would likely flow from metaphors drawn from our experience of having a body. Thus, they depart from the tradition of truth-conditional semantics, but rather view meaning in terms of conceptualization. Meaning is viewed in terms of mental spaces instead of in terms of models of the world.

3.3 Language as Embodied Information

Based on the discussion both in chapter 2 and in preceding parts of this chapter we hypothesize that natural language faculty can be thought of as a being a mechanism for conveying embodied information and that language itself, both its generic structural features and specific utterances, is such information. What do we mean by embodied information? The defining characteristics of this abstract concept are that it is contextual by definition and meaningful in practice. Batesons [14] definition of information as a difference that makes a difference is a related concept. In other words, it is the kind of information a system in the aforementioned fifth system class (an open system with feedback, energy management and reproduction) processes when it operates towards fulfilling its goals. The use of natural language by humans is contextual and pragmatically meaningful. Furthermore, as proposed by cognitive linguistics, the structure of language is fundamentally grounded on the characteristics (enabling and constraining in terms of survival and reproduction) of the environment we operate in. Embodied information processing is further discussed in context of language in [15].

4 Conclusion

In this chapter we first overviewed the state of art when it comes to the foundations of natural language and how it relates to the forms of communication used by other animals. Much work has been put into developing theories which would explain the structure of language and how it relates to the information processing capabilities of the human mind. Second, we presented a hierarchy of generic information processing systems and situated natural language in that hierarchy. Based on this we arrived at viewing, on a philosophical level, human language faculty as an embodied information processing system in the tradition of the embodied theory of mind.

References

1. Gardner, T., Naef, F., Nottebohm, F.: Freedom and Rules: The Acquisition and Reprogramming of a Bird's Learned Song Science. Science 308, 1046–1049 (2005)
2. Hauser, M.D., Chomsky, N., Fitch, W.T.: The Faculty of Language: What Is It, Who Has It, and How Did It Evolve? Science 298, 1569–1579 (2002)
3. Hauser, M.D., Fitch, W.T.: Computational Constraints on Syntactic Processing in a Nonhuman Primate. Science 303, 377–380 (2004)
4. Friederici, A.D., Bahlmann, J., Heim, S., Schubotz, R.I., Anwander, A.: The brain differentiates human and non-human grammars: Functional localization and structural connectivity. Proc. Natl. Acad. Sci. USA 103, 2458–2463 (2006)
5. Pinker, S., Jackendoff, R.: The faculty of language: what's special about it? Cognition 95, 201–236 (2005)
6. Gentner, T.Q., Fenn, K.M., Margoliash, D., Nusbaum, H.C.: Recursive syntactic pattern learning by songbirds. Nature 440, 1204–1207 (2006)
7. Premack, D.: Is language the key to human intelligence? Science 303, 318–320 (2004)
8. Kirby, S., Dowman, M., Griffiths, T.L.: Innateness and culture in the evolution of language. Proc. Natl. Acad. Sci. USA 104, 5241–5245 (2007)
9. Shannon, C.E.: A Mathematical Theory of Communication. Bell System Technical Journal 27, 379–423, 623–656 (1948)
10. Wilson, M.: Six Views of Embodied Cognition. Psychonomic Bulletin & Review 9, 625–636 (2002)
11. Lakoff, G., Johnson, M.: Philosophy In The Flesh: the Embodied Mind and its Challenge to Western Thought. Basic Books (1999)
12. Croft, W., Curse, D.A.: Cognitive Linguistics. Cambridge University Press, Cambridge (2004)
13. Johnson, M.: The Body in the Mind: The Bodily Basis of Meaning, Imagination, and Reason. University of Chicago Press, Chicago (1987)
14. Bateson, G.: Steps to an ecology of mind. University Of Chicago Press, Chicago (2000)
15. Feldman, J.: From Molecule to Metaphor. The MIT Press, Cambridge (2006)

Author Index

Aldana, Maximino   113

Biddle, Fred G.   65

Charlebois, Daniel A.   89
Cloud, Daniel   9

Eales, Brenda A.   65
Emmert-Streib, Frank   179

Feyt, Rolando Pajon   29

Gingras, Carey   29

Hahne, Lauri   153

Jacob, Christian   29

Kaern, Mads   89
Kesseli, Juha   153

Lloyd-Price, Jason   195

Martínez-Mekler, Gustavo   113

Niiranen, Samuli   1, 219
Nykter, Matti   153

Perkins, Theodore J.   89

Ribeiro, Andre   1
Ribeiro, Andre S.   195

Sarpe, Vladimir   29
Shmulevich, Ilya   153

Yli-Harja, Olli   153, 219
Yli-Hietanen, Jari   219

Zañudo, Jorge G.T.   113

DATE DUE